MAPPING BIOLOGY KNOWLEDGE

Science & Technology Education Library

VOLUME 11

SCOPE

The book series *Science & Technology Education Library* provides a publication forum for scholarship in science and technology education. It aims to publish innovative books which are at the forefront of the field. Monographs as well as collections of papers will be published.

Mapping Biology Knowledge

by

KATHLEEN M. FISHER
San Diego State University, U.S.A.

JAMES H. WANDERSEE
Louisiana State University, U.S.A.

and

DAVID E. MOODY
San Diego State University, U.S.A.

KLUWER ACADEMIC PUBLISHERS
DORDRECHT / BOSTON / LONDON

A C.I.P. Ctalogue record for this book is available from the Library of Congress.

ISBN 1-4020-0273-4
Transferred to Digital Print 2001

Published by Kluwer Academic Publishers,
P.O. Box 17, 3300 AA Dordrecht, The Netherlands.

Sold and distributed in North, Central and South America
by Kluwer Academic Publishers,
101 Philip Drive, Norwell, MA 02061, U.S.A.

In all other countries, sold and distributed
by Kluwer Academic Publishers,
P.O. Box 322, 3300 AH Dordrecht, The Netherlands.

Printed on acid-free paper

CONTENTS

Introduction   1
Kathleen M. Fischer

Chapter 1   Overview of Knowledge Mapping   5
Kathleen M. Fischer

Chapter 2   The Nature of Biology Knowledge   25
James H. Wandersee, Kathleen M. Fischer & David E. Moody

Chapter 3   Knowing Biology   39
James H. Wandersee & Kathleen M. Fischer

Chapter 4   Student Misconceptions in Biology   55
Kathleen M. Fischer & David E. Moody

Chapter 5   Meaningful and Mindful Learning   77
Kathleen M. Fischer

Chapter 6   Language, Analogy, and Biology   95
James H. Wandersee

Chapter 7   Using Concept Circle Diagramming as a Knowledge Mapping Tool   109
James H. Wandersee

Chapter 8   Using Concept Mapping as a Knowledge Mapping Tool   127
James H. Wandersee

Chapter 9   SemNet® Semantic Networking   143
Kathleen M. Fischer

Chapter 10 The Paradox of the Textbook   167
David E. Moody

References   185

Author Index   201

Subject Index   207

KATHLEEN M. FISHER

INTRODUCTION

Mapping Biology Knowledge addresses two key topics in the context of biology, meaningful learning and the role of knowledge mapping in promoting meaningful learning. Chapter 1 provides an overview of several common strategies for spatial knowledge representation, Chapters 2–6 discuss some of the key considerations in learning for understanding, and Chapters 7–10 describe several metacognitive mapping tools and the research that informs their use.

Please note that the chapters are written in different voices and thus have different styles, tones and ways of referring to the authors, depending upon the particular authorship of each chapter. A brief description of the chapters is given below.

Road maps are regularly used by travelers on land, sailors use their charts when they go to sea, and scientists often rely on spatial knowledge maps when they practice science. Science maps range from the long-established periodic table (now available in many delightful and useful forms as internet hypertext documents) to biochemical pathways to the newer human genome maps. Likewise, semantic or word-based knowledge maps are often used by students, teachers and researchers as learning, teaching, knowledge navigation, and assessment tools. *Chapter 1, Introduction to Knowledge Mapping* by Fisher, provides an overview of word-based knowledge mapping including concept maps, cluster maps, webs, semantic networks, and conceptual graphs.

The domain of biology is vast, the depth of knowledge in many areas is awesome, and the knowledge structure of the field is both complex and irregular. In addition, biology knowledge is assimilated from many different sources, both formal and informal. For these and perhaps other reasons, knowledge mapping seems to be particularly useful for those interested in mastering biology. These issues are examined in *Chapter 2, The Nature of Biology Knowledge*, by Wandersee, Fisher and Moody. This chapter also considers the "two cultures" influencing biology education, scientists and science educators.

In many biology courses, students become so mired in detail that they fail to grasp the big picture. Overemphasis on detail accounts in part for the fact that relatively few Americans understand how trees "construct themselves from thin air" (Schnepps, 1997b), even though nearly all have studied photosynthesis at least once and often several times. Yet memorizing trivial detail is not a goal of science learning. A more useful approach is for the learner to construct a well-ordered overview of the big ideas and their interrelations, combined with skill in knowing how to find more information as needed. *Chapter 3, Knowing Biology* by Wandersee and Fisher, describes a little-known *system analysis* of biology as one example of a high-level

1

overview (Miller, 1978). It presents the human mind as an expectation-generator that will hold onto information it perceives valuable for anticipating future events and that will discard information it perceives as irrelevant. The "need to know" principle can be helpful in deciding the level of detail students must have in a given situation.

It is now well established that students' minds are not blank slates and that students' preconceptions or naive conceptions can present major impediments to learning. This is especially true in a field like biology where there is a lot of folk knowledge and personal experience. *Chapter 4, Student Misconceptions in Biology* by Fisher and Moody, reviews this widely researched phenomenon.

Meaning-making is achieved in part through mindful learning, the use of fluid and flexible thinking. *Chapter 5, Meaningful and Mindful Learning* by Fisher, reviews Langer's (1989, 1997) seven myths of education, including ideas such as overdrilling (rote learning) and "work now, play later." This chapter prompts teachers to examine their commitment to "coverage" of "facts" at the cost of meaning-making and development of thinking skills.

Most of our thoughts lie below the surface of conscious awareness, just as most of an iceberg is submerged beneath the sea. And just as only the tips of icebergs are visible to us, so only the tips of our thoughts are available to conscious knowing. And to carry the analogy one step further, just as an iceberg sunk the unsinkable *Titanic*, so subconscious thoughts can sink or at least subvert a lesson. This is the topic of *Chapter 6, Language, Analogy, and Biology* by Wandersee. This chapter concludes the examination of meaning-making, looking at how biology jargon and analogies can help or hinder understanding.

Metacognitive tools serve as support systems for the mind, creating an arena in which we can make our knowledge explicit, reflect on its organization, and polish its edges. These tools are also useful for building and assessing students' content and cognitive skills. Concept circle diagrams are metacognitive tools that can help students build their skills in categorizing, which is essential to constructing knowledge hierarchies and to learning complex information. This topic is presented in *Chapter 7, Using Concept Circle Diagramming as a Knowledge Mapping Tool* by Wandersee.

If you want to see where you have been and where you are going, get a map. This advice is as basic for students learning science as it is for travelers on the road. *Chapter 8, Using Concept Mapping as a Knowledge Mapping Tool* by Wandersee, describes Novakian concept maps. The chapter is organized using Frequently Asked Questions (FAQs).

Ideally, just as we can look into a mirror to see our faces, so it would be nice to gaze into a clever machine to examine our minds. Unfortunately, this clever machine has yet to be developed. However, the SemNet® software provides a crude approximation, allowing us to see explicitly see how we and our students think about a given topic at a given point in time. *Chapter 9, SemNet® Semantic Networking* by Fisher, provides an overview of the SemNet® tool in the classroom.

Textbooks are integral components in biology teaching and learning. Mapping tools can be used by readers to increase their access to the content of a text and by writers and other interested people to analyze the structure of a text. *Chapter 10, The*

Paradox of the Textbook by Moody, provides a historical overview of biology texts and illustrates one approach to analyzing the importance of a topic, in this case evolution, in texts over time.

KATHLEEN M. FISHER

CHAPTER 1

Overview of Knowledge Mapping

If You Want to Find Your Way, Get a Map!

Sara and Charlotte, driving from Cincinnati to San Francisco, leave the freeway in Colorado and soon realize they are lost. Sara, who is driving, asks Charlotte to get out the map so they can find their way again.

Susan and Roy, exploring the islands of the Caribbean in a Catamaran, get blown off course by a storm and aren't sure where they are. They take a reading on the GPS and pull out a chart to find their location.

Adam and Paul, taking a biochemistry course in college, find themselves hopelessly lost in the voluminous new material. They sit down over a weekend and map out where they have been and where they are going in the course, and return on Monday in much firmer control of their destiny.

WHAT IS KNOWLEDGE MAPPING?

Knowledge mapping or knowledge representation is a process in which a schematic representation of knowledge is created. Knowledge maps typically include the most important concepts (usually noun ideas) in boxes, ovals, or circles (Figure 1.1). Concepts are usually connected by lines which are often unlabeled (and thus represent mere associations, as in "is somehow related to") and are sometimes given name labels. When the lines (links, relations, arcs) are labeled, it is usually with a verb phrase. The relationship indicated by a line between two concepts is always bidirectional, but the name label that is shown on a map may be either unidirectional or bidirectional. Arrowheads are often included on the line so the reader knows which way the relation should be read, but in hierarchical maps, arrowheads are often omitted on the assumption that the reader will read from top to bottom.

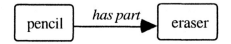

Figure 1.1. Elements of knowledge mapping include concepts such as <u>pencil</u> and <u>eraser</u>, links such as "has part", and propositions such as "<u>pencil</u> has part <u>eraser</u>".

It appears that knowledge mapping has originated independently multiple times and in multiple contexts. As one example, a young woman who recently worked for me had invented knowledge mapping on her own, as a tool for learning. To the best of her knowledge, she had never heard of or seen knowledge maps created by others. Her maps were hand drawn in rich colors, similar to the Mind Maps and Visual Thinking Networks described briefly below. Additional discoveries of knowledge mapping are described below.

A BRIEF HISTORY OF KNOWLEDGE MAPPING

Knowledge mapping began early, when cave men and women sketched their knowledge about their environment in the form of symbols on the walls of caves. We'll skip much history between these early events and modern times. The history presented below makes no effort to be comprehensive, but instead captures some of the highlights of knowledge mapping in education of particular interest to us.

According to Brachman and Levesque (1985), knowledge representation as a means of creating artificial intelligence (AI) in computers began in the 1950s. Specifically, they cite a 1950 paper by Turing and Shannon (1950) and a conference at Dartmouth in 1956 as the starting points for serious work in AI. The goal of AI is concerned with "writing down descriptions of the world in such a way that an intelligent machine can come to new conclusions about its environment by formally manipulating these descriptions (Brachman and Levesque, 1985, p. xiii). AI requires much more elaborate mapping techniques than those desirable in education.

The goals of knowledge mapping in education are quite different from those in AI. Educational knowledge mapping is seen primarily as a tool for learning, teaching, research, intellectual analysis, and as a means for organizing knowledge resources. In all fields using knowledge mapping, the idea is to tap into the workings of the brain. AI and education are two sides of a coin. AI wants to use knowledge mapping to build computers that mimic the brain's intelligence, while educators want to use knowledge mapping to stimulate and support students' efforts to increase their intelligent use of their own innate resources.

Gordon Pask developed many forms of cybernetic knowledge mapping in the 1950s through the 1970s, during which he published at least three books and 150 papers. His interest in mapping was applied to studies involving such topics as the "Styles and strategies of learning" (Pask, 1976a) and "Conversational techniques in the study and practice of education" (Pask, 1976b). He developed maps to represent the ideas that emerged in student conversations and to show the connections between those ideas (Pask, 1975, 1977). Since researchers today are once again turning to discourse and dialogic analysis, it seems likely that they will also find knowledge mapping helpful.

Pask straddled the worlds of AI and education, as indicated by his dual appointments as Professor in the Department of Cybernetics at Brunel University and Professor in the Institute of Educational Technology at the Open University, both in Great Britain. These two topics are combined in his 1975 book, *Conversation, cognition and learning: A cybernetic theory and methodology*. In the introduction,

Pask describes his theory as being concerned with psychological, linguistic, epistemological, ethological, and social mental events of which there is awareness – that is, conscious thoughts and interactions. He was obviously ahead of his time, at least in education. But researchers today might appreciate the many strategies he developed for mapping the dynamics of a conversation.

In the same decade but on a different continent, science educator Novak and his graduate students invented concept mapping as a learning tool for K–12 students (Stewart, Van Kirk, & Rowell, 1979). Novakian concept maps grew out of Ausubelian learning theory (1963, 1968) with its emphasis on building connections between ideas. Novakian concept maps (described further in Chapter 8) are widely used in science teaching today from elementary school through the university.

With the advent of the Macintosh personal computer in the early 1980s, Fisher, Faletti and their colleagues created the SemNet® knowledge mapping software as a learning tool for college biology students (Fisher, Faletti, Patterson. Thornton, Lipson & Spring, 1987, 1990). The major objective was to help students shift from their prevailing rote learning methods to meaningful understanding of biology content. The design of this software grew directly out of AI and cognitive science, especially Quillian's (1967, 1968, 1969) semantic network theory for how we store information in long term memory (see Chapter 9 for more information).

Also in the 1980s, Wandersee (1987) developed concept circle diagrams (CCDs) for the purpose of helping students clarify their thinking about inclusive/ exclusive relationships. Being able to organize ideas into categories and to distinguish between similar but different things are key steps in learning and are supported by the use of CCDs (discussed in Chapter 7).

In the late 1980s and early 90s, Horn (1989) in the US and Buzon (e.g., Buzon & Buzon, 1993) in Great Britain took knowledge mapping into the business world. In fact, Buzon has been a tireless promoter of his strategy, Mind Mapping, in both education and business arenas throughout the British Empire. Buzon is interested in mapping as a means of promoting creativity and divergent thinking, and has developed the MindMan software to support his style of mapping (Table 1.1). Probably the best commercial success in knowledge mapping, at least in the US, is the Inspiration software (Table 1.1), a concept mapping tool available for both IBM and Macintosh platforms.

In the late 1990s we have witnessed the amazing growth and blossoming of the World Wide Web. The quantity of information available at our fingertips is staggering, and the need for intelligent, user-friendly mapping strategies grows stronger every day. So far, this need has not been adequately answered, although various efforts are being made (see, for example, Table 1.1).

Table 1.1. Some knowledge mapping software described on the internet, 1999

Software	World Wide Web Site
The Axon Idea Processor	http://web.singnet.com.sg/~axon2000/article.htm
Banxia Software	http://www.banxia.co.uk/banxia/
CoCo Systems Limited	http://www.coco.co.uk/
Inxight Hyperbolic Trees	http://www.inxight.com/Content/7.html
Inspiration Software	http://www.inspiration.com/
LifeMap	http://www2.ucsc.edu/mlrg/lifemapusermanual375/lifemapusermanual375.html
MindMan Software	http://www.mindman.com/
SemioMap Builder	http://www.semio.com/download/Download.cgi
SemNet Software	http://trumpet.sdsu.edu/semnet.html
Smart Ideas	http://www.smarttech.com/smartideas.htm
VisiMap	http://www.coco.co.uk/prodvm.html

HOW DOES KNOWLEDGE MAPPING HELP STUDENTS LEARN?

Research suggests that in more cases than not, knowledge mapping exercises of all types help students learn. Why is this? There are many possible answers to this question. First, mapping provides sustained support for *time on task* in thinking about a topic. Second, if mapping is done collaboratively, it can lead to *extended discussions about the meanings of concepts and the relations between them*. Third, the act of creating an organized structure of ideas on paper or in a computer necessitates and often *prompts the creation of such a knowledge structure in the mind*. Fourth, knowledge mapping prompts students to *take implicit, often fuzzy, associations and make them into explicit and precise linkages*, a process that is at the heart of meaning-making. Fifth, knowledge mapping *takes many cognitive and metacognitive skills that remained invisible for so many generations and makes them visible, explicit, and accessible.* Sixth, mapping prompts students to *make finer discriminations between ideas*, another process at the heart of learning. Seventh, the more one practices, *the better one becomes at organizing and relating concepts* (Cliburn, 1990). And eighth, each time two concepts are joined with a relation in working memory, *that information is believed to be "broadcast" to all the modules in the brain* so it can be used to solve any current problem the vast subconscious brain may be working on (Baars, 1988).

Jonassen, Beissner, & Yacci (1993, p. 8–10) describe the advantages of knowledge mapping in another way. First, they say, semantic structure is inherent in all knowledge. Second, structural (organized, semantic) knowledge is essential for recall and comprehension. Third, learners assimilate structural knowledge effectively. Fourth, knowledge structures in memory reflect the world. Fifth, structural knowledge is essential to problem solving. And sixth, there are significant differences between the structural knowledge of novices and experts, so that for novices, working on their structural knowledge to make it more expert-like is a natural part of learning.

HOW CAN KNOWLEDGE MAPPING CONTRIBUTE
TO EDUCATIONAL REFORM?

Mapping is a tool for personal and social knowledge construction and a tool that supports meaningful learning. In the classroom, mapping can provide
- structure for the *minds-on* part of *hands-on/minds-on* teaching,
- a systematic means for reflecting on and analyzing inquiry learning,
- a knowledge arena for operating on ideas, and
- tangible support for the transition from teacher-centered to student-centered classrooms.

WHAT IS THE EDUCATIONAL REFORM MOVEMENT?

Serious educational reform began in the 1970s in Great Britain and Australia. In the early 1980s the US came on board. The momentum of reform has steadily gathered steam ever since.

The reform movement advocates meaningful science learning at every grade level. The group in the American Association for the Advancement of Science (AAAS) that is working toward reform is called Project 2061, to signify their expectation that it will take that long (until the year 2061) to revamp education in the US. AAAS has produced several well-known guidelines to help the process along (1983, 1989, 1998), and has succeeded in bringing the two cultures (scientists and science educators) together to work on the project. The National Research Council (1996) also has taken a leadership role, as have many other professional and granting agencies.

Among other things, reform documents (Appendix 1.1) repeatedly cite the need for strategies that help science learners acquire interconnectivity and discrimination among science ideas, two features that most clearly differentiate novices from experts and most dramatically affect recall and application of knowledge. It also happens that these two features especially benefit from knowledge mapping activities. Cohen (1991, p. 46), in studying a newly reformed mathematics classroom, describes the problem succinctly:

> If the recent reforms are to succeed, students and teachers must not simply absorb a new body of knowledge. Rather, they must acquire a new way of thinking about knowledge and a new practice of acquiring it. They must cultivate new strategies of problem solving that seem to be quite unusual among adult Americans. They must learn to treat knowledge as something they construct, test and explore, rather than as something they accumulate.

One obstacle to achieving reform is that many teachers are confused or overwhelmed by the demands of teaching science for understanding (Flick, 1997). They understandably mix many of their old teaching strategies with the new (Cohen, 1991). Further, American schools have not been organized to support continued growth and learning by teachers (although this is changing slowly and in piecemeal ways). Teachers lack the basic requirements of a professional workplace such as a work station and telephone, and they are not given work time to prepare their lessons or to collaborate with and observe their colleagues. It is perhaps because of issues such as these that, while inquiry learning and learning for understanding have been promoted several times during this century (Cain & Evans, 1979; Pask, 1975; Woodburn & Obourn, 1965), they have repeatedly suffered lack of long term success in American public schools.

We believe that today's reform has a greater chance of being successful because it is strongly supported in at least four ways. First, it is informed by a quarter century of research on learning and teaching (e.g., Wittrock, 1974a, 1974b; Norman, Rumelhart, & the LNR Research Group, 1975; Thro, 1978; Sowa, 1983; Novak & Gowin, 1984; West, Fensham, & Garrard, 1985; Salomon & Globerson, 1987; Langer, 1997).

Second, the current reform movement has generated important policy decisions and resources at international, national, state and local levels (see Appendix 1-1 for a selected list of relevant documents).

Third, many of the skills involved in meaningful science learning have been identified and explicated (e.g., Donaldson, 1978; Resnick, 1987a; Driver, Squires, Rushworth, & Wood-Robinson, 1994; Langer, 1997).

And fourth, the new electronic technologies, when properly harnessed, can significantly enhance the learning process and change the nature of the classroom (e.g., Papert, 1980; Clements & Gullo, 1984; Crovello, 1984; Pea, & Gomes, 1992). In fact, cognitive research has provided reasonably strong direction for appropriate ways of using technology (including knowledge mapping) to support the acquisition of cognitive and metacognitive skills for learning science and mathematics (e.g., Pea, 1985; Perkins & Saloman, 1989; Pea & Gomes, 1992; Jonassen, Beissner, & Yacci, 1993; Fisher & Kibby, 1996).

We expect to see continued growth and development of knowledge mapping techniques for educational and commercial purposes, in part because humans have always valued maps of their domains, and in part because electronic technologies make so many new strategies possible. Below we briefly describe some of the mapping strategies you will be encountering in this book, in order of increasing complexity.

METACOGNITIVE TOOLS FOR KNOWLEDGE MAPPING

A. Cluster Maps and Webs

Clustering and webbing are techniques that capture associations between ideas. Cluster mapping was developed as a creative writing technique by Rico (1983). The idea is that clusters are produced by "first gaining access to the natural functions of the right brain and its predilection for wholeness, images, and metaphors, followed by a conscious collaboration with the syntactical, logical left brain" (Ambron, 1988, p. 122). Ambron (1988) uses the clustering technique in teaching college biology to nonmajors. Students work in collaborative groups to produce maps. According to the author, this approach increases students' active participation in learning and reduces their anxiety. Figure 1.2 reproduces a cluster map from Ambron's (1988) paper.

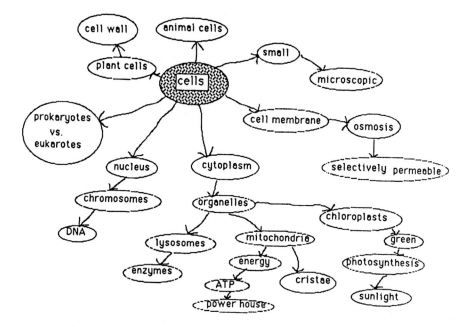

Figure 1.2. Cluster map of the concept, <u>cells</u>, from Ambron (1988, p. 123). Reproduced with permission from the Journal of College Science Teaching.

Webbing is a similar technique that aims specifically to build a bridge between students' prior knowledge and new ideas. Lovitt and Burk (1988) introduced webbing into Farmers Branch Elementary School in Dallas, Texas, where more than half the students were from minority ethnic groups. Twenty-two of 32 teachers in the school used the technique the first year. Student scores in language improved in all grades, and scores in reading improved in grades 1, 3 and 6. Science learning was not assessed in this study.

Figure 1.3 illustrates webbing. Webbing can be more complex than shown here, in that detail can be added about each of the related concepts. Students in Farmers Elementary drew their webs with pencil and paper, while our adaptation below was done with MacDraw software. Figure 1.3 also illustrates the complexity of information children typically have about a single familiar concept such as <u>animals</u>.

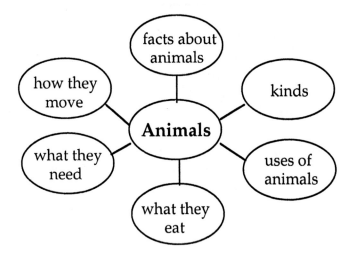

Figure 1.3. Sample web adapted from Lovitt & Burk (1988, p. 120).

B. Mind Maps

Mind Maps® (Figure 1.4) are similar to cluster maps and webs, but they have been developed and promoted independently by Tony Buzon in Great Britain (Buzon & Buzon, 1993). Buzon sees Mind Maps as a way to capture and reflect processes in the brain. In Buzon's words, "Each individual brain cell is capable of contacting and embracing as many as 10,000 or more proximate brain cells in the same instant. It is in these shimmering and incessant embraces that the infinite patterns, the infinite maps of your Mind, are created, nurtured and grown. Radiant Thinking reflects your internal structure and processes. The Mind Map is your external mirror of your own Radiant Thinking and allows you access into this vast thinking powerhouse" (Buzon & Buzon, 1993, p. 31). An example of a Buzon-like Mind Map is shown in Figure 1.4.

Buzon has enthusiastically promoted the Mind Map across Europe and the far reaches of the former British Empire, emphasizing it as a means of increasing creativity and performance. Mind Mapping is also useful for capturing and providing an overview of a complex set of ideas (Figure 1.5). MindMan software for creating Mind Maps electronically is available for IBM compatible platforms (Table 1.1).

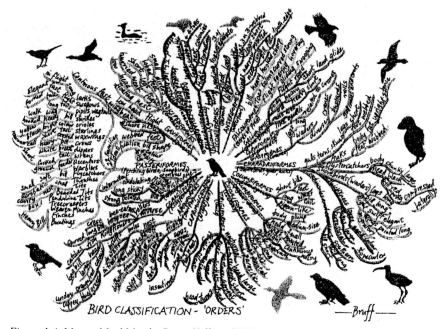

Figure 1.4. Master Mind Map by Brian Heller of IBM summarizing a lifetime study of birds, from Buzon & Buzon, 1993, p. 243. Reprinted with permission from Plume Books.

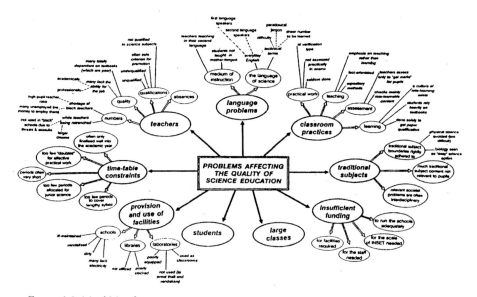

Figure 1.5. Mind Map from a paper entitled "Problems and issues in science education in South Africa" (Sanders, 1992).

C. Computer-Generated Associative Networks

Cluster maps, webs, and Mind Maps are strategies for people to use to help them think about and learn a topic. When used as learning tools they can be considered input devices. Mapping can also be used to see what is already inside a person's mind. In this case, mapping becomes an output device.

One approach for assessing how people think about a topic is to ask them to generate relatedness ratings for pairs of concepts. For example, indicate how closely the following pairs of terms are related on a scale from 1 (not at all) to 5 (highly related):

bear — mammal
mammal — whale
reptile — snake
horse — reptile

The goal is to capture the perceived similarities and distances between ideas and represent them graphically. Schvaneveldt (1990) developed the Pathfinder software to do this. The end product looks a bit like a cluster map or web (Figure 1.6), but it is important to recognize that the method of generation is different and therefore what can be inferred from it is also different. To repeat, Pathfinder is not a tool to support learning but rather a tool for measuring learning that has occurred previously. The other mapping procedures we describe here can be used in a similar way, to measure learning output, but in this context we have been describing them primarily in the context of learning (input).

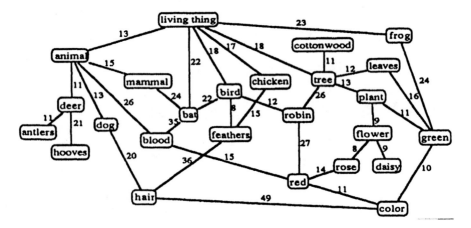

Figure 1.6. Pathfinder associative network about living things (Branaghan, 1990, p.114).

D. Concept Circle Diagrams

As noted above, concept circle diagrams (CCDs) were developed by Wandersee (1987). These diagrams help students to understand inclusive/exclusive relations among elements and categories. Figure 1.7 illustrates a concept circle diagram describing seed plants.

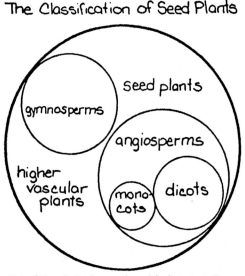

Note: Colored shading on the original student diagram made the individual concepts more distinct.

Figure 1.7. The classification of seed plants, from Wandersee (1987, p. 16), with permission from Science Activities.

Constructing CCDs can be useful starting point for students, helping them to build their skills for categorizing and hierarchy construction prior to creating concept maps or other knowledge maps. To add detail, any concept in a CCD can be telescoped (Figure 1.8).

E. Concept Maps

Concept mapping was invented by Joseph Novak and several of his graduate students as described above and has since been enthusiastically promoted by Novak (e.g., Novak, 1964, 1991a, 1991b; Novak & Gowin, 1984). It is probably the most widely used method of knowledge representation in science education in the US.

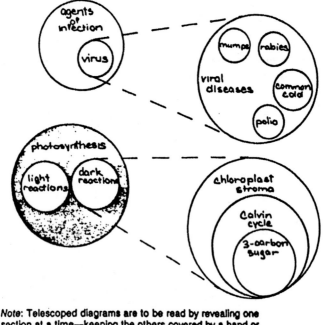

Note: Telescoped diagrams are to be read by revealing one
section at a time—keeping the others covered by a hand or
piece of blank paper. Reading is done by moving from the
left diagram to the right diagram. The explanatory sentence
that accompanies each section should be read right after
viewing it.

*Figure 1.8. Using telescoping to elaborate in concept circle diagrams, from Wandersee (1987,
p. 16), with permission from* Science Activities.

Concept mapping has grown out of Ausubelian learning theory (Ausubel, Novak, &
Hanesian, 1968) and is consistent with constructivist theory (Kelly, 1955; Wittrock,
1974a, 1974b; Pope, 1982; Pope & Gilbert, 1983). Both theories emphasize the
importance of connections between ideas and of personal, individual knowledge
construction by each learner. A concept map by J. Wandersee describing entomology
is shown in Figure 1.9.

Concept mapping is a useful tool for curriculum analysis and planning (Starr &
Krajcik, 1990). Concept maps can also effectively capture the growth in concept
understanding over time, as shown in such studies as "A twelve-year study of science
concept learning" by Novak and Musonda (1991) and "Tracing conceptual change in
preservice teachers" by Morine-Dershimer (1993). Likewise, there is some evidence
that concept maps can help teachers become more effective (Beyerbach & Smith,
1990; Hoz, Tomer & Tamir, 1990).

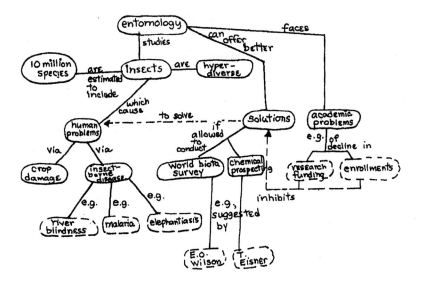

Figure 1.9. Concept map about entomology, by James H. Wandersee, summarizing an article in Science *magazine.*

A meta-analysis of nineteen studies of student learning with concept mapping found that mapping has medium positive effects on student achievement and large positive effects on student attitudes (Horton, McConney, Woods, Senn, & Hamelin, 1993). The gains are especially evident in biology courses. As one example of an individual study, Okebukola (1990) examined student performance in a biology course with 138 biology majors. The 63 students who used concept mapping to study genetics and ecology significantly outperformed the others on a test for meaningful learning of genetics (discussed further in Chapter 8).

F. Semantic Networks

The SemNet® software for the Macintosh was initially designed as a learning tool for use by students, especially those in college biology classrooms (Fisher, Faletti, Patterson, Thornton, Lipson & Spring, 1987, 1990), although it now enjoys much wider use (Fisher, 1990; Fisher & Kibby, 1996). SemNet® provides a model for the way in which denotative factual information is organized and functions in long-term memory. The software allows individuals to construct large networks of ideas containing dozens or hundreds or thousands of concepts. Bidirectional links are formed between pairs of related concepts. A well-developed concept in a semantic network may have as many as thirty links to other concepts. Nets can be viewed one frame at a time (Figure 1.10). Many additional views (about 20) of the knowledge structure are provided by the software, but really good high level maps of such complex structures have yet to be created.

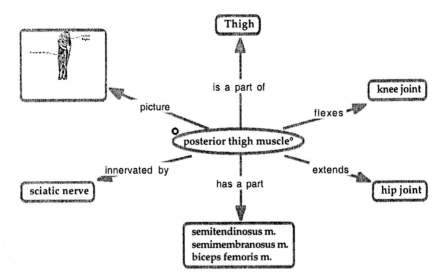

Figure 1.10. A frame from a semantic network about human anatomy created by Dr. Hugh Patterson at the University of California—San Francisco Medical School. The frame contains one central concept, <u>posterior thigh muscle</u>, with its links to seven related concepts. It also contains a picture that can be enlarged by double-clicking.

Because semantic networking is not something people have consciously done in the past, Joseph Lipson (one of the developers of SemNet®) predicted that we will only see the full benefit of SemNet® when a significant cadre of people has practiced for years and become as fluent in creating and interpreting semantic networks as they now are in reading and writing. SemNet® semantic networking is described further in Chapter 9.

G. Conceptual Graphs

So far we have discussed knowledge mapping as a learning tool and assessment tool. The conceptual graphs introduced here have been used primarily as a research tool. Conceptual graphs (Figure 1.11) differ from concept maps and semantic networks in that concept nodes may contain concepts, events, states, goals, and other elements, and these can be described by simple names or complex propositions. The nodes are connected by named unidirectional or bidirectional relations. Overall, complexity rises to a higher level. Conceptual graphs are usually laid out on large paper maps but have sometimes been created with a special version of the SemNet® software that does not limit the number of characters in a concept node (the usual limitation is 31 characters).

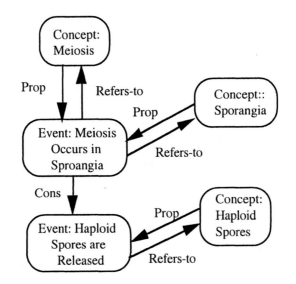

Figure 1.11. Conceptual graph segment for meiosis adapted from Gordon (1996, p. 216).
Relation abbreviations are: Cons = Consequence, Prop = Property.

Conceptual graphs are especially useful for eliciting knowledge from experts (Gordon, 1989, 1996). Gordon (1996) has found that a small finite number of relations can be used successfully to elicit knowledge from experts in many fields. These are the relations on the lines. The expert can introduce any number of new relations as she or he describes the subject. These relations are classified by type (concepts, events, states, goals, etc.) and placed in the nodes.

H. Visual Thinking Networking (VTN)

The Visual Thinking Network (VTN) is a new technique being developed by Palma J. Longo (1999). It incorporates many of the features of Mind Mapping, including color, shapes, graphics, and "playfulness" in representations, but also adds named unidirectional and bidirectional names to the links between ideas.

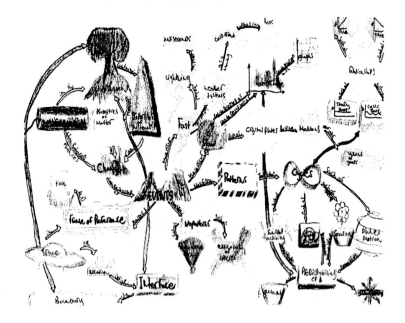

Figure 1.12. Visual Thinking Network created by a student to represent beliefs about the earth and the seasons. (Longo, 1999, p. 3).

I. Summary of Mapping Styles

Table 1.2 summarizes some of the key differences among the mapping techniques described in this chapter. This table includes the context or purpose for each type of mapping described in the accompanying source(s). However, it is important to realize that mapping is a flexible, adaptable tool, and nearly all forms of knowledge mapping have been adapted in many different ways and for many different purposes. The table also identifies the key elements usually used to construct each type of map, although again, this can vary.

Table 1.2. Modes of Knowledge Mapping Used in Education. (Table developed collaboratively by Kathleen M. Fisher, SDSU; Robert Abrams, University of California—Santa Cruz; Palma Longo, Columbia University; and James H. Wandersee, Louisiana State University.)

Source	Knowledge Mapping Form	Unit of Knowledge Construction	Nature of Link	Purpose
Ambron, J. (1988). 'Clustering: An interactive technique to enhance learning in biology', Journal of College Science Teaching (18), 122–144.	Cluster Maps	concept	unlabeled association	stimulate creative writing
Lovitt, Z. & Burk, J. (1988). 'Webbing: A bridge between teaching and learning', Educational Horizons (66), 119–121.	Webs	concept	unlabeled association	link prior to current knowledge
Buzon, T. & Buzon, B. (1993). The mind map book: How to use radiant thinking to maximize your brain's untapped potential, New York, Plume Book (Penguin).	Mind Maps	concept	unlabeled association	stimulate creativity
Schvaneveldt, R.W. (1990). 'Proximities, networks, and schemata', in R. Schvaneveldt (ed.), Pathfinder associative networks: Studies in knowledge organization, Norwood, NJ, Ablex, 135–148.	Computer-Generated Associative Networks	concept pairs	unlabeled association	capture concept pair similarity and distance
Wandersee, J.H. (1987). 'Drawing concept circles: A new way to teach and test students', Science Activities 24(4), 1, 9–20.	Concept Circle Diagrams	concept label, concept circle, text	color-coded inclusive, exclusive and overlapping circles	cluster concepts & clarify categorical relations

Table 1.2 continued. Modes of Knowledge Mapping Used in Education.

Concept Maps	concept (instance)	unlabeled association or unidirectional link	capture concept microstructure or topic macrostructure (or both)	Novak, J.D. (1964). 'The importance of conceptual schemes for science teaching', *The Science Teacher* 31(6), 10. Novak, J. & Gowin, D.B. (1984). *Learning how to learn*, Cambridge, UK, Cambridge University Press.
Semantic Networks	instance	bidirectional link	capture concept microstructure *and* topic macrostructure	Fisher, K.M., Faletti, J., Patterson, H.A., Thornton, R., Lipson, J. & Spring, C. (1990). 'Computer-based concept mapping: SemNet software — A tool for describing knowledge networks', *Journal of College Science Teaching* (19), 347–352. Fisher, K.M. (1990). 'Semantic networking: The new kid on the block', *Journal of Research in Science Teaching* (27), 1001–1018.
Conceptual Graph	proposition	constrained bidirectional links	capture expert knowledge	Gordon, S.E. (1989). 'Theory and methods for knowledge acquisition', *AI Applications* 3(3), 19–30. Gordon, S.E. (1996). 'Eliciting and representing biology knowledge with conceptual graph structures', in K.M. Fisher & M.R. Kibby (eds.), *Knowledge acquisition, organization, and use in biology* (NATO ASI Series F, Vol. 148), New York, Springer Verlag, 135–154.
Visual Thinking Strategies (VTN)	color, form, concept labels, spatial information	unidirectional links, bidirectional links	help students make their knowledge explicit	Longo, P. (in preparation). 'The influence of visual thinking networking promoting long term meaningful learning and achievement for 9th grade earth science students', Thesis, Columbia University.

Mapping of the geographic terrain has been under development for several centuries. Improvements have been steadily added, so that geographic mapping is now quite sophisticated. The advent of satellites and the enormous volumes of information they collect have challenged geographic mapping to move into new territory. Fortunately, computers are able to help manage the complexity arising from multiple ways of seeing (infrared, visible spectrum, etc.), multiple scales and perspectives, and multiple layers of information about a single location.

Knowledge mapping in contrast is a very new field, barely 50 years old in its modern incarnation. It is a field that is still in search of the right metaphors, algorithms, and conventions. The need for good effective mapping strategies grows with each new day of the knowledge explosion.

WHAT DO MAPS AND ASSOCIATIVE NETWORKS REPRESENT?

According to schema theory, knowledge is stored in our minds in the form of mental constructs of ideas (Rumelhart, 1980; Rumelhart & Ortony, 1977). These mental constructs are schemas for different types of things. Each schema has a variety of slots or relations to other ideas. Two very general schemas in biology are those for objects and processes. For example, the generalized schema for an object would seek answers to the following questions: What is it part of? What parts does it have? What are its characteristics? What is its function? Where is it found? How big is it? Each of these questions would become a relation that points to the answer. In contrast, the generalized schema for a process might seek answers to these questions: What is its function? What are its inputs or substrates? What does it produce? Under what conditions does it occur? What are its active agents? Where does it occur?

Three basic premises of schema theory are that (a) human memory is organized semantically and schematically, (b) schemas consist of organizations of interrelated concepts which often have name labels, and (c) abstract templates (schema) are subconsciously extracted from particular experiences.

There is a high degree of correspondence between external knowledge maps created by an individual and internal knowledge structures in that individual's long term memory. This correspondence is to the *conceptual* organization of an individual's ideas, *not* necessarily to the physical organization of their mind. The correspondence is tempered by the dynamic nature of the mind that can change perspectives instantaneously. A single perspective is captured in a knowledge map.

A knowledge map might be likened to a snapshot of a runner in a race. The snapshot captures one aspect of the individual's posture and form at one moment in time, among the many variations that could be seen in the entire running event. Another way to think about it is that while knowledge maps can capture some aspects of the idea structures in an individual's long-term memory, they more directly reflect recent events in the individual's short term or working memory.

Knowledge maps not only reflect the structural knowledge of an individual, but can also promote and capture changes in that knowledge. By serving as an extension of working memory and an arena in which individuals can operate on ideas, knowledge mapping is an effective learning process.

JAMES H. WANDERSEE, KATHLEEN M. FISHER, &
DAVID E. MOODY

CHAPTER 2

The Nature of Biology Knowledge

Genetic Drift?

Twelve faculty members in a major university's genetics department were collaborating
to produce a televised genetics course. Unexpectedly, they discovered that they were
unable to agree on the definition of a gene — the basic unit of their field. How could that
be? With so many experts and a rather well-understood entity, how could there be so
much dissension about what a gene is?

WHAT IS INVOLVED IN "KNOWING BIOLOGY?"

The problem that the geneticists in the opening vignette were facing is not necessarily
an unusual situation. Although an outsider might be startled by it, insiders will not be.
A biologist's biology knowledge, like all knowledge, consists of various kinds of
mental representations — declarative and procedural, logical and emotional,
experiential and received, private and public, semantic and structural, basic and
applied. The twelve geneticists each had their own specialization, so each knew
different parts of the genetics subdomain. Each one viewed genetics through the lens
of his own preparation, experience, and specialization. Each had also learned his
genetics at different times, under different conditions, and in different ways.
Disagreements such as this are generally more common among specialists than
among generalists – in part because·of the details associated with learning a particular
subfield in depth, and in part because of the experts' deep emotional attachment to
their own hard-won views of the subject matter.

Not only do specialists view their subject through different lenses, but a study by
one of the authors (Abrams & Wandersee, 1995) finds that expert ideas about biology
knowledge change over time. Beginning biologists typically believe that biology
knowledge is derived solely from observations of the living world, as shown in
Figure 2.1.

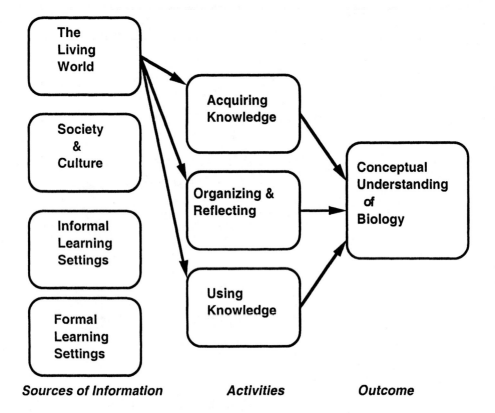

Figure 2.1. Naive view of sources of information influencing biology knowledge. Adapted from Wandersee, 1996.

Gradually, however, practicing biologists come to realize that what we already know affects how we see, acquire, organize, and use new biological knowledge, and even how we perceive and interact with the living world (Figure 2.2).

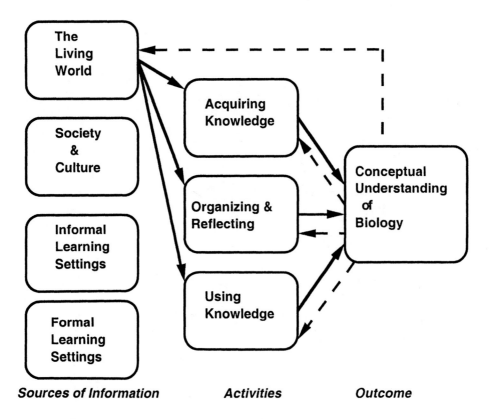

Figure 2.2. Recognition that what we know about biology influences what we see.

But even this is not the whole picture. While the living world is obviously the most pertinent source of biological information, our conceptual understandings of life (and the conceptual understandings of our students) are inevitably influenced by secondary sources such as society, culture, informal learning, and the theoretical constructs we derive from our formal learning (Figure 2.3). Our basic assumptions about what is likely, what is possible, and what is impossible are derived from attitudes and values we develop from these background sources over a lifetime. This is called one's worldview following Joseph I. Lipson (1980, personal communication; see also Cobern, 1996). Worldview may enable individuals to be receptive to certain new ideas or closed to them.

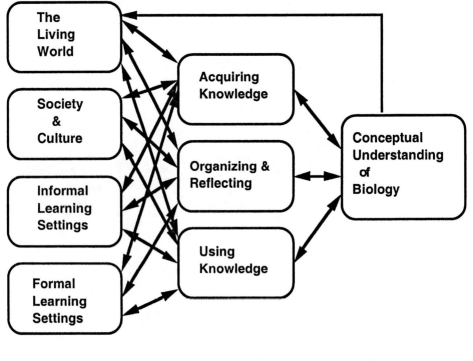

Figure 2.3. A more sophisticated view of learning: Recognition that our cultural assumptions, metaphors for understanding, and overall worldview influence what we see. Modified from Wandersee, 1996.

THE NATURE OF BIOLOGY

The word "biology" originated in the 19th century. The precursors of this broad field of study were natural history, botany, and medicine, including anatomy and physiology (Mayr, 1982, p. 36). Darwin's theoretical and evidentiary legacy, coupled with Mendel's work, unified all of biology and established its explanatory power (Atrans, 1990). Molecular biology and the "modern synthesis" extend the powers of explanation and in some cases allow prediction. Neo-Darwinian evolutionary thought informs our understanding of ultimate causality, while detailed elucidation of DNA gives valuable insights into the proximal causes of cellular control as well as elucidation of phylogenetic relationships both currently and throughout evolutionary time.

Biology is the study of living things. But what are living things? Is all life cellular as claimed by the cell theory? Or are our intimate noncellular parasites, the viruses, also alive as some argue? And if so, are prions alive? And when does human life begin? At conception? At the onset of neurophysiological activity? At birth? Or is life simply continuous, passed on from cell to cell? Is a person actually dead when her brain stops functioning or when her heart stops beating? Likewise, is the tomato we've just picked up from the grocery store alive? If we slice that tomato, is the slice alive? If we take a seed from that slice, is it alive? And is a cluster of naturally root-grafted White Pine trees really a single super-organism (Kourik, 1997)?

Defining life is not a simple task. Life's boundaries remain much "fuzzier" than we'd like in spite of (or perhaps because of) the many recent advances in our knowledge. Its fuzzy edges are just one of the ways in which the life sciences differ from the physical sciences. The contrasts range from the nature of the objects and events being studied to what counts as an explanation. The form and content of theory and the generalizability of explanations are also significantly different in the life sciences (Rosenberg, 1985, p. 34). Thus, like Ernst Mayr (1982), we respect the physical sciences but do not aspire to become them. Biology is a gradually maturing science, but that does not mean it is simply on its way to becoming more like physics and chemistry. As one example, the levels of organization that characterize life on earth (atom, molecule, cell, tissue, organ, organ system, organism, population, community, ecosystem, biome, and biosphere), each with its own emergent properties, have no close parallels in the physical sciences.

Biologists study objects that have (and vary in) information content and whose history matters, whereas chemists typically study inanimate objects such as atoms and molecules that are essentially interchangeable. Life consists of open systems of a certain minimum complexity. These systems self-regulate, self-repair, maintain a steady state, develop, reproduce, and are seriously constrained by their requirements for survival (Miller, 1973, p. 69). The dynamic, synthesizing, organizing, energy-consuming nature of living things sets them apart from inanimate objects.

John Moore holds that evolution, genetics, and developmental biology are "the core of conceptual biology," because these subdomains focus on "the fundamental characteristic of life — its ability to replicate over time (1993, p. viii)." From the organism's genetic program, to the differentiation that occurs as a single cell becomes a multicellular organism, to the enhancement of the survival of the species that natural selection confers, life has ancestry that cannot be ignored.

Bronowski's Rule (Bronowski, & Mazlish, 1960, p. 218) claims that confidence in any science is proportional to the degree to which it is made mathematical. This rule may be appropriate for the physical sciences, but is not broadly applicable in biology, even though some subfields (e.g., cell physiology, genetics, ecology) make use of mathematics. Another difference is that "The objects with which physical science deals do not have goals, ends, purposes, or functions (except as they serve explicit human purposes)" (Rosenberg, 1985, p. 43). Limb buds in a chick embryo, in contrast, do have developmental goals programmed into their DNA. Under normal circumstances, limb buds consistently and eventually become wings. For these and other reasons, the living and nonliving worlds are profoundly different.

In summary, biology is a unique science, quite different from the physical sciences. Biology knowledge is extensive, highly complex, incomplete, and often ill-structured. The domain of biology stretches across great expanses of time and a remarkable array of subfields. The breadth and complexity of biology, the interconnectedness of knowledge at many different levels, and the invisible nature of many key processes make biology a particularly difficult subject to teach and to learn.

TEACHING BIOLOGY

Biology includes literally hundreds of subdomains. These subdomains differ widely with respect to terms, methods, goals, and issues. The field of biology is so large that no one person can possibly know all of it. For example, the Human Genome Project is currently constructing the analog of a periodic table for biology in the form of a series of genetic maps, not only for humans but also for Drosophila, yeast, a worm, a virus and a plant. This project alone has such scope and magnitude that a person could devote a lifetime to understanding it fully.

Because of the scope and complexity of the subject, the biology teacher forever walks a tightrope between comprehensibility and misrepresentation. Ausubel pointed out that learning is a series of diminishing deceptions (Ausubel, Novak, & Hanesian, 1978). As our knowledge of a domain grows, we are continually learning the limitations of, boundaries for, and exceptions to what we learned earlier. However, one cannot leap directly from *novice* status to that of *expert*. It takes an extended series of small learning jumps to get there. There is nothing inherently wrong with teaching and learning a simplified version of what professional scientists know – as long as both teacher and student are aware that this version is simplified in specific ways.

Biology and its subdomains have developed many specialized concepts and relations (Faletti & Fisher, 1995). Therefore, mapping biology or its subdomains is no easy task. However, cartography and understanding go hand in hand (Wandersee, 1990). We submit that creating scientifically valid maps of the cognitive territory can enhance progress in biology learning and in biological research. It is an especially valuable tool for learners — when the students do the mapping.

We are indeed pleased to read that "biology education is flourishing at many campuses as new approaches to teaching and learning take hold" (Howard Hughes Medical Institute, 1996, p. 3), but we wonder whether or not these new approaches include *organizing and reflecting upon the biological knowledge being acquired* – coprocesses we consider crucial to meaningful learning in biology (see Figure 2.3) and quite different from the generic thinking skills being promoted by others.

This book focuses on mapping biology knowledge. Consider the value of mapping. Just as, in nature, ants lay down a pheromone trail that marks the way to a food source and persists as long is it is renewed and used, so do we need to mark our cognitive progress and regularly revisit important thinking pathways. Otherwise, the antagonistic but necessary process of forgetting will make those pathways fade and vanish like ant trails.

KEY POINTS UNDERLYING OUR VIEWS ON BIOLOGY MAPPING

Knowing Biology

1. Knowledge Sources. Biology knowledge is acquired from many sources and is shaped by our existing conceptions. It is not derived only from observations of the living world.

2. Ways of Knowing. Different kinds of biologists have different ways of studying and thinking about biology.

3. Fluid and Flexible Use. The better one's biology knowledge is organized, the better it can be fluidly and flexibly used in the real world.

4. Learning. There is a gradual progression of biological understanding from novice (little/no formal biology knowledge) to advanced novice (some formal biology knowledge) to specialist (considerable formal biology knowledge) to expert (in-depth knowledge of an area of biology), and it takes a lot of work to move through this progression. The best we can hope for in any initial course in a subfield of biology (e.g., Introduction to Botany) is to move the student from novice to advanced novice.

Biology Knowledge

1. Integration. Biology knowledge includes both integrated networks of ideas and isolated packets of knowledge. Concepts are elaborated in memory by such things as images, examples, and analogies.

2. Vastness. Biology knowledge today is vast, ever evolving, somewhat ill-structured, and dependent upon context and method.

3. Fuzzy Edges. The edges of biology, and of life itself, are fuzzy, ill-defined, and problematic.

4. Subdomains. Biology consists of a large number of subdomains that range from the molecular to the interstellar (e.g., life on Mars), from the present to the distant past, from the marine to the terrestrial and beyond, from the autotrophic to the parasitic, from the submicroscopic to the macroscopic, and from the minute and precellular to the huge and multicellular.

5. Uniqueness. Biology makes use of, but differs in significant ways from, the physical sciences. Biology is not on its way to becoming physics, nor should we expect it to be.

Biology Instruction

What may be called "rigorous instruction" in biology is often inappropriate, reflecting the shape of expert knowledge rather than taking into account the state of the learner and the path that must be traveled in order to understand biology in a meaningful and mindful way.

THE VALUE OF MAPPING

Biology can be summarized in the form of general principles that do not require many biological terms, as has been done by Hoagland and Dodson (1995). Some biologists see only the inherent limitations of such statements, but principle sets such as the one above can form an effective scaffold for further learning. General principles offer an important initial colonization structure on which to build a more comprehensive biological understanding.

In learning about and traversing such a complex territory as biology, the metaphor of a map is useful. Mapping biology can help us see where we (and others before us) have been, and plot a route to our new destination via established referents, landmarks, and benchmarks. The alternative is accidental accretion of disorganized knowledge by random walk. However, if you want to get where you want to go, use a map!

Yet, at the same time, we must always recognize that any map has inherent limitations. The map is not the territory. There are always unplotted, alternative paths to a destination. Consider a college campus. Although there are formal concrete and asphalt walkways linking all of the buildings, these are not the only trajectories that pedestrians use. You will also find what landscape architects call "desire lines" – "tracks carved into the grass by those who don't follow the paths" (Muschamp, 1997, p. A16). Desire lines graphically represent unanticipated human needs such as: minimizing distance, time, effort, and temperature change. That's why the most effective maps follow rather than precede, human experience.

AN AMERICAN TALE OF BIOLOGICAL ILLITERACY

A basic goal of contemporary biology instruction is to produce students who can be described as *biologically literate* (American Association for the Advancement of Science, 1983, 1989, 1998). The intent is that biologically literate students will be capable of valid biological thinking as citizens, as personal and public decision-makers, and as employees in the global economic network. In contrast, consider this true story of biological illiteracy.

The Farmer's Museum in Cooperstown, New York, contains what was purported to be a 10-foot petrified man — the subject of the Cardiff Giant hoax. In 1869, a former archaeology and paleontology student named George Hull hired some sculptors in Chicago to create his giant from a slab of gypsum (having blue venation) that Hull had quarried in Fort Dodge, Iowa. The resulting anatomically accurate sculpture, prematurely "aged" using sulfuric acid and ink, was then shipped by rail to the farm of Hull's cousin, William Newell, located near Cardiff, New York. It was secretly buried in the middle of the night, between Newell's barn and house, and left untouched for half a year.

When Hull thought the timing was right (because some million-year-old fossil bones had recently been dug up on a farm near Newell's), he directed his cousin to hire two well diggers and tell them exactly where he wanted a new well dug (above the buried giant). Then Newell waited until the two men rushed in to say that, while digging his well, they had encountered the body of a giant-turned-to-stone.

As the conspirators had hoped, the news about the amazing discovery spread quickly, and by mid-afternoon Newell pitched a tent over the excavation and began charging 25 cents per person for admission. Soon the price was raised to 50 cents, as visitors came by the thousands. Experts' opinions were split regarding whether it was an ancient statue or a fossilized human giant (perhaps one of the giants to which the biblical Book of Genesis referred). No one said it was a fake.

In 10 days, the find began receiving national attention. Hull sold a two-thirds interest in the giant to a syndicate who moved it to an exhibit hall in Syracuse and began charging a $1 admission fee. From there, it was exhibited in various other cities including New York City. P. T. Barnum made the syndicate an offer of $50,000 to buy the giant. When his offer was refused, Barnum had a stone copy of the giant carved for him, and then claimed his competitors had sold him their original giant and were currently exhibiting a fake! Thousands now rushed to see Barnum's giant, since he was a master of publicity.

Finally, a leading paleontologist from Yale University, Professor Marsh, examined the original giant and pronounced it to be of very recent origin. Only then did the public begin to come to its senses. Barnum admitted his giant was a fake, but noted it was only a fake of a fake. On his death bed, Hull gave a full account of the whole affair. He and Barnum, independently, had profited grandly from the ignorance of the masses and apparently neither felt any shame in doing so.

While the American public may be less gullible in matters biological today, those of us charged with teaching biology ought not be complacent. For example, there are many health care scams, in part because Americans are frequently in *biological denial*. Many think that diet pills can keep them thin, that the effects of aging can be lastingly reversed by cosmetics or surgery, and that cigar smoking entails no cancer risk. Others assume that the earth can sustain as many inhabitants as humans can produce, that nonhuman species are expendable if economics calls for their demise, and that science and medicine can conquer any human disease. Some people believe that evolution is unimportant or even antithetical to understanding life on earth, that the earth's oceans are a vast and inexhaustible resource and a bottomless garbage dump, that calories don't count, and that plants eat soil. Of all such weaknesses, the failure to understand population dynamics, biogeochemical cycles, and sustainability are the most ominous and urgent problems of biology education today.

SCIENCE TEACHING AND ITS "TWO CULTURES"

For decades, there have existed two cultures of science teaching – that of the science educators (e.g., those affiliated with graduate schools of education and working with the public schools) and that of the scientists (e.g., those in colleges of basic or natural sciences). Further, there is a widespread tendency to attribute all the failures of contemporary science instruction to whichever of the two groups dominates science curriculum design and testing at the moment.

Science educators complain that curricula designed primarily by the scientists embody unrealistic expectations about what students can and want to learn at a given grade level. Scientists complain that they *are* the content experts and are thus better judges of what constitutes sound science learning. They lament that the US is falling behind other nations in science learning based on selected international test results. They ascribe this state of affairs to the "watering down" of K–12 science courses by the faddish "educational establishment" that they see as ignorant of the fundamentals of science. Such a critique may end with the scientists uncharacteristically invoking a conspiracy theory (involving teacher unions, laziness, greed, and the perceived erroneous notion that science learning should be fun — not work) to account for it (Gross, 1997). Mostly the two cultures remain ignorant about and disinterested in each another.

While members of both cultures have earned research-based Ph.D.s, the science educators have specialized in studying the teaching and learning of science while the scientists have specialized in advancing science knowledge. As a result, one camp has a greater knowledge of human cognition while the other has a greater knowledge of science content. History shows that responsibility for science curriculum design and science testing tends to oscillate between the two cultures as political conditions change and one group is brought in to remedy the perceived "ill effects" of the other.

Typifying this tension is a *New York Times* article by University of Virginia emeritus professor of biology, Paul R. Gross (1997, p. 21) entitled "Science Without Scientists." The article refers to "...the long history of philosophical skirmishes between those in the public education establishment (including those in the graduate schools of education) and the academics who generally reside in university arts and sciences departments."

The article centers on what the author sees as a ridiculous decision of the State of California to hire "a group made up mostly of professional educators based at California State College at San Bernardino" as consultants for establishing new science guidelines when they could have used the services of a group of scientists (including Nobel laureates) to perform the work for free. Since the full set of facts about the State's hiring decision is not included, the reader must rely upon the author's interpretation of events.

The author's interpretation is undermined, however, by his attempt to portray the other side as comprised primarily of incompetent agents with ulterior motives and a disdain for science. The article is peppered with invective like "public education bureaucrats," "the same bureaucratic foolishness that gave us ebonics," "Nobel laureates aren't good enough for California's curriculum study," "teachers' bureaucracies," "exclusionary rules," "self-righteous isolationism," and "further debasement...[of] public understanding of science."

The driving force behind the article appears to be this statement: "Thus your typical working scientist considers the quality of science education in the public schools to be dismal, a judgment for which there is solid evidence." (Gross, 1997, p. 21). Indeed, the Associated Scientists group involved in the controversy wrote in its original proposal, "'The dumbing down' of science in school curricula has prevented most productive dialogue between scientists and K–12 teachers, and is a continuing source of frustration and disappointment" (Triangle Coalition, 1997). When average citizens read such an article in a reputable newspaper, it is no wonder if they later vote against increasing local school funding. Competitive cycles of this sort make the prophecy of doom for US science education self-fulfilling.

How can the US escape this negative force field? Clearly the country has to move beyond the two cultures with their culturally biased views of what science education should be and also move beyond their recurrent media jousting. It is disheartening to find scientists who claim to value research and yet disregard decades of research on science teaching and learning. And it is equally disheartening to find science education researchers who consistently avoid contact with science and scientists. Scientists must become aware that their involvement in improving science education cannot be sporadic or predicated on their intellectual superiority if it is to have any significant positive effect. Science education researchers must realize that they have chosen to work at the interface between science and society, and thus need input from both and owe accountability to both (Wandersee, 1983). Both cultures must realize that they can be strong and have great impact by working together.

Until the two groups work together, US science education will continue to careen from reform cycle to reform cycle. The American Association for the Advancement of Science's Project 2061 (1983, 1989, 1998), which brings the two groups together, is a much needed breath of fresh air. We hope that Project 2061 will stay on course and continue to receive funding for the long-term effort it promises to make — building upon the advice of scientists, science education researchers, and practicing science teachers. We also hope this type of collaboration will become the rule rather than the exception.

Scientists and science education researchers working in collaboration can identify those scientific ideas that are most important to teach and learn, can design a coherent curriculum that lets each new idea build logically upon previous ones, can design instructional experiences that make these important ideas accessible to a wide range of students, and can provide students with intellectual challenges that will help them reach their potential. Most importantly, the two cultures can teach one another, learn from one another, build essential bridges, and present a united front to the establishment.

We all need to understand that: (a) there is no quick-fix in the real world of public schools; (b) science education can benefit from the long-term involvement of scientists; (c) science teachers' science knowledge may lack depth and conceptual integration, but can be striking in its breadth; and (d) new lab activities, new curricula, new textbooks, and new technologies alone are not the answer. Poor instruction, for example, can neutralize the effects of a fine curriculum and vice versa.

Those of us who work outside the schools also need to temper our expectations about what the nation's science teachers ought to accomplish by witnessing firsthand the physical, social, and economic realities of the *lower tier* of schools. These are the schools that most teachers and students would leave immediately if they weren't trapped there by circumstances. It isn't impossible to teach and learn science there, but it is several orders of magnitude more difficult. Overcrowded classes, noisy hallways, disrespect for knowledge and authority, physical discomforts (too hot, too cold, too wet), police presence, minimal science equipment, minimal science budget, gang activity, nonexistent laboratory facilities, chronically absent students, ancient textbooks, dysfunctional restrooms, outdated computers, leaking roofs, and apathetic parents each can detract from quality science education. When many of these factors are combined in a single environment, all education becomes compromised. Cultural influences like gangsta rap, MTV, the fashion industry, professional sports, and underground economies further contribute to the erosion of a good learning environment. Science education may be significantly improved by "fixing" these and other problems, but these fixes go beyond the two cultures: this is a job for society.

BIOLOGY OF THE BRAIN HAS LITTLE TO TEACH US ABOUT HOW WE SHOULD TEACH BIOLOGY

In a thoughtful analysis of the contemporary relationship between neuroscience and education, Bruer (1997) points out that while brain science fascinates scientists and teachers alike, it actually has little to offer educational practice. In Bruer's own words, "Currently, we do not know enough about brain development and neural function to link that understanding directly, in any meaningful, defensible way, to instruction and educational practice (p. 4)." For example, teaching to a particular hemisphere of the brain, using brain-based curricula, teaching to a preferred learning style, and synaptic change-based early childhood instruction are popular in some areas but are not supported by neuroscience findings.

Bruer (1997, p. 10) demonstrates that cognitive psychology, which studies the nature of mind and the mental functions that underlie observed human behavior, holds a justifiable claim to be the "basic science of learning and teaching." Cognitive science is a bridge to both education and to cognitive neuroscience. Our emerging biological knowledge of the human brain may someday contribute to educational practice, but we will have to be able to link brain structures to cognitive functions to instructional goals and desired outcomes. Cognitive psychology's studies of expertise, knowledge acquisition and representation, alternative conceptions in science, and analogical thought currently yield greater insights for educational improvement than do PET scans or studies of neurons and synapses. "In the meantime," Bruer writes, "we should remain skeptical about brain-based educational practice and policy" (p. 15).

AN INVITATION TO INQUIRY

In this chapter we have roughly circumscribed the boundaries of our interests and perspectives regarding biology learning. There is, of course, much more to say about the nature of biology. We will be introducing more ideas as you progress through the book, using the "just in time" strategy so popular in today's industrial world.

Our primary message is that we have reason to think that the process and the products of mapping biology knowledge can help learners organize and integrate their biological knowledge in systematic, coherent, and meaningful ways. Such biology maps can support classroom discussion, as well as reflection upon and revision of the learner's growing knowledge base. For each hour of active engagement in biological observation or experiment, we think roughly two hours of focused reflection are required to make scientific sense of it — to realize its educational yield, so to speak. Mapping can serve as such a focused, reflective strategy.

Now, we invite you, the reader, to accompany us on a journey of intellectual exploration — to see whether or not mapping biology knowledge is actually a fruitful learning approach that you can use for understanding and remembering how life works.

JAMES H. WANDERSEE & KATHLEEN M. FISHER

CHAPTER 3

Knowing Biology

Is Blood Type Related to One's Character?

In contemporary Japan, knowing a person's blood type is not just considered important during blood transfusions, it is also used to predict an individual's personality and the nature of his or her social interactions (Sakurai, 1997). Young people who go out on a first date typically try to learn each other's blood type—or, ask their own matchmaker to screen out the undesired types in advance. Employers in Tokyo may seek to hire only employees who have a particular blood type—one socially compatible with the employees they already have. People who read the Japanese tabloids hope to discover what blood types their favorite TV and film stars have. Women's magazines even publish diets said to be suitable for particular blood types. In Japan, the subject of blood types is as popular a topic of general conversation as the weather.

From the history of biology, we know that many so-called common sense ideas have turned out to be erroneous when subjected to the light of careful scientific scrutiny. Human blood, for example, has been attributed with having extraordinary powers far beyond the role we ascribe to it today as a physiological fluid in the form of a liquid tissue – with past claims including that it acts as the seat of the soul, as the prime determinant of human inheritance, and as the controlling agent of human personality.

With respect to the latter, Hippocrates promoted blood-letting methods to adjust human personality characteristics using the doctrine of the four humors (Gardner, 1972, p. 58). Thomas Bartholin (1616–1680) "reported that he had examined a young girl who displayed feline characteristics after drinking the blood of a cat" (Magner, 1979, p. 116). Even that giant of biological thought, Charles Darwin, proposed a blood-borne theory of inheritance in which tiny gemmules that were given off by every body cell were carried to the reproductive organs and assembled into eggs or sperm (Magner, 1979, pp. 409–410). Darwin thought that, at conception, blood-borne gemmules arising from both parents formed the new human embryo – with gemmules for particular traits coming from either the maternal or paternal line.

While the possibility always exists that blood-based explanations of human personality may someday prevail in science, their future looks bleak at this juncture. From what we know about inheritance of personality today, claiming linkage patterns between ABO blood type genes and personality-influencing genes seems far-fetched as a comprehensive explanation. Some proponents claim that the very fact that today's science rejects their views only substantiates how progressive their views really are. However, as the popular scientist Carl Sagan (1979, p. 64) pointed out in his writings about borderline science, while it is true that people sometimes laugh at those whose thinking is actually far more advanced than their own, such laughing alone is not convincing validation – since people also laugh at Bozo the Clown, and rightly so.

The arbiter in science is convincing and replicable evidence. Until it exists, speculation must be treated as speculation. The big contribution which a scientific theory makes is bringing order out of chaotic facts and observations; while the ABO blood-type theory of personality does that to some extent – it must also fit with the biological knowledge we currently have about human blood and about human personality's heritable and environmental components. Social science tells us that personality differences go well beyond biologically defined temperaments. Prevailing moods may reflect long-term positive or negative experiences – they may derive from each individual's personal and social learning history within particular familial or cultural contexts (Snow, Corno, & Jackson, 1996, p. 258). In short, human personality determination is apparently quite complex and has multiple causes. Perhaps it is well to recall Alfred North Whitehead's oft-quoted aphorism, "Seek simplicity, but distrust it."

The idea that ABO blood type influences personality dates back to 1930 when, during Japan's Asian military invasions, the Japanese military commissioned a study of how blood type affects personality — in an attempt to create better soldiers. Some proponents have sought anthropological data to support these claims. Yet there seems to be little scientific evidence to support the conclusion that ABO blood type influences personality and few other cultures share this belief. Is it fact or fantasy?

The idea still persists in Japan today – across all age segments of the population. The Japanese believe that type A blood (the "farmers" type) produces nervous, detail-oriented, honest, loyal, careful accommodators; type B blood (the "hunters" type) produces noisy, proud, aggressive, optimistic, adventurous people; type AB blood (the "humanists" type) produces creative, critical but useful people who are full of contradictions; and type O blood (the "warriors" type) produces highly motivated, workaholic, emotional people who seek to control any group they join.

What can we learn about blood types from biology? In 1901, Karl Landsteiner reported that there were types of human blood that together constituted the ABO blood system, and that the incompatibility of certain blood types could explain the rapid intravascular hemolysis that occurs during some blood transfusions. In 1930, he received the Nobel Prize in medicine for his discovery of human blood groups. The ABO blood types are produced by a single gene for which there are three different alleles (variations of the gene) in the population. These alleles produce enzymes that modify carbohydrates attached to the surface of red blood cells. The carbohydrates are antigenic — that is, they can stimulate production of antibodies and will react with antibodies that are specific to them. Today, we know that there are many other blood group antigens on red blood cells, the most important of the others being the Rhesus (Rh) system. There is also a complex set of antigens (the HLO antigens) on white blood cells and many other body cells.

As for the ABO system, we now know that the A, B, and O factors are carbohydrates (oligosaccharides) that attach to the ceramide lipids of the red blood cell's plasma membrane, but that can also attach to proteins. Type O cells are marked with a saccharide sequence — fucose-galactose-n-acetylglucosamine-galactose-glucose — attached to the ceramide membrane lipid. The A antigen is produced when N-acetylgalactosamine is attached to the outer galactose in this sugar, while the B antigen is produced when an extra galactose is attached to that outer galactose. Thus, what humans inherit is either 1) an allele coding for an enzyme that attaches N-acetylglactosamine to the O saccharide (type A), 2) an allele coding for an enzyme that attaches an extra galactose to the O saccharide (type B), 3) both of these alleles (type AB), or 4) two alleles that do not alter the basic saccharide (type O) — see Figure 3.1.

Humans produce antibodies that circulate in the blood and that react with type A and type B antigens. A type A person will have anti-B antibodies, a type B person has anti-A antibodies, and a type O person has both kinds of antibodies. Interestingly, these antibodies are produced in response to antigens found on intestinal bacteria, but react with type A and type B red blood cells due to cross-reactivity. The antibodies are not produced if an antigen is part of "self." Most people do not make antibodies that react with the type O saccharide. Recently, however, some rare individuals have been discovered who do produce antibodies that react with the O saccharide — the Bombay phenotype.

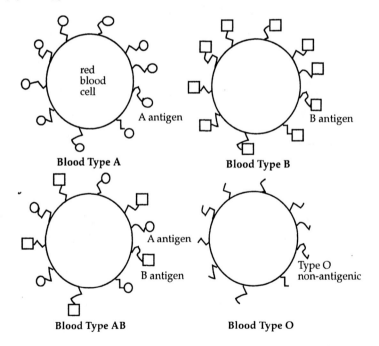

Figure 3.1. Schematic diagram of glycolipids on the surface of red blood cells that are produced by ABO alleles and give rise to the ABO blood types. Drawing by Laura Becvar.

What does this ABO cell biology have to do, if anything, with human character determination? Seemingly nothing. Science is mute on this point, and there is virtually no scientific evidence to support an "ABO personality hypothesis." It is interesting to note that the Japanese blood type study was mandated in the same year that Karl Landsteiner won the Nobel prize, which may be why this particular red blood cell system (ABO) was selected as the scientific tool to use for human personality prediction. Yet while blood typing *is* scientific, such simplistic and unwarranted leaps of application definitely are not.

Humans seem to have strong desire to predict personality — it is part of their future orientation. In the US astrology serves this purpose, while in Japan the ABO blood system is used. Science cannot support either approach because there is no theoretical basis, no known mechanism, and questionable empirical data.

On the other hand, scientists must always reserve final judgment. Consider the recent National Institutes of Health findings showing success in treating certain medical conditions using the traditional Chinese therapeutic technique of acupuncture, or the recent Baylor College of Medicine pilot study showing that magnetic therapy (using small, 300- to 500-gauss magnets fitted to the anatomic area where the pain is centered) successfully reduces pain in patients suffering from post-polio syndrome (Altman, 1997). Both therapies initially seemed dubious to scientists, and unfortunately they still don't understand the scientific basis for these therapeutic effects. Right now, two leading hypotheses for the magnetic therapy include the following: the magnets may increase blood flow to a painful area of the body – reducing inflammation and pain, or, the magnetic field may effectively block pain receptors in the painful area (Fremerman, 1998, p. 56). These therapies contrast with many other popular remedies for medical conditions that have been shown to be ineffective.

Such topics are not typically the foci of scientific research because scientists are more likely to make progress via studies that are supported by and have the potential to advance sound scientific theories. Scientists are justifiably reluctant to work on investigations in the so-called *borderline* or *fringe areas* of science. They are willing to pass on studies with a low probability (albeit, potentially high yield) for success, those that require hypotheses which cannot be supported by current scientific theory. The Japanese blood type theory of character determination and the popular astrological approaches to forecasting human events fall into this category. Today the scientific research topics being pursued are determined mostly by where the funding is available, but since scientists are involved in establishing the funding programs, the same biases still apply, albeit indirectly.

HUMANS SEEK TO INTEGRATE THEIR KNOWLEDGE FOR FUTURE USE

The foregoing story illustrates that advanced societies expect science to be able to explain everything—even social behavior. But, science has its limits (both as to what constitutes a legitimate scientific question and as to what is currently explainable scientifically). Science doesn't have all the answers and never will. It is likely that individual human behavior will always remain unpredictable to some degree. The leap from basic biology to behavior is enormously challenging, in that it entails many levels of biological organization, environmental factors, and the effects of learning from experience.

In spite of these reservations, we agree with psychoanalyst George Kelly (1955, p. 48) who maintains that humans ultimately seek to anticipate real events. Such anticipation is crucial for survival of the individual and the species. Humankind is future-focused. In fact, Kelly says that humans are "tantalized" by the future and this is why we argue that humans' knowledge structures reflect this bias.

People search for recurrent events and the conditions under which they occur. The relations humans use to connect the concepts that they have already learned serve primarily to represent reality for future reference and application; relations make possible the conceptual hierarchies that serve to "rank-order" and integrate what we know for efficient use later. Dennett (1996, p. 57) puts it this way, "A mind is fundamentally an anticipator, an expectation-generator." The process of knowledge mapping is useful in this regard in that it helps us to make our relations explicit and to streamline our knowledge structures for ease of retrieval.

THE IMPORTANCE OF THE "NEED TO KNOW" PRINCIPLE

It appears that some organisms have little need to know things in advance. The amoeba does not seem to have a plan or even a focused "search image" of what it must seek out or avoid. It responds to selected stimuli "on the fly." An economy-of-information rule seemingly applies across the kingdoms of life—although the quality and quantity of what needs to be known in advance varies with the species. Thus, each extant species of organism has, over time, developed perceptual and representational limits adequate for its survival to date.

This is not necessarily so for contemporary humans. As "informavores," we have, quite recently in our history, been led to think that more information is always better than less. Unfortunately, such a superabundant stimulus flux can also lead to what has been called "information overload" and "paralysis of analysis."

We suggest that in biology teaching and learning, students' knowledge structures should be optimized primarily for efficiency and effectiveness in making anticipatory decisions. Many complex details that probably will not be used frequently in the near future can be "off-loaded" to external memory devices (e.g., books, computer storage devices, or visual media). Dennett (1996, p. 134) points out that such off-loading can free us from the processing limitations of our brain—which is far from the largest in the animal kingdom — thus, streamlining our thinking.

Biology teachers have traditionally foisted high volume/high conceptual density memorization tasks upon their students—claiming these to be a requirement for "understanding" biology and an indicator of their courses' high academic rigor. (We think that knowledge-mapping tasks would be a better alternative to such assignments—more on this later.) And while these fact-laden assignments are usually not solely rote-memorization tasks, they do tend to induce a high level of rote memorization. Few instructors would want to ask a former graduate to retake her final exam five years hence in, for instance, plant physiology, to see what course-based knowledge is still accessible today. While many students, especially biology majors, are able to memorize and reconstruct selected biology topics in great detail within the context of a particular biology course, biology teachers are generally aware how little of that information each student stores in long-term memory. And the quantity of long-term understanding declines precipitously with nonmajors.

SPURIOUS CRITICISM OF STUDENTS' UNDERSTANDING

Craig (1997, p. 23), in a short essay on how woefully inadequate today's University of Michigan students' knowledge of American historical and political knowledge is, exemplifies the carping of those university professors who apparently have not thought through which knowledge in their field is of greatest worth.

He relates that every semester for the past 10 years, he has given his undergraduate classes on public opinion, consisting mostly of upper class students, a "brief quiz of assorted historical facts." Later he dubs these facts to be "basic historical and political knowledge."

What is this foundational knowledge the "bright, inquisitive individuals" in his class lack? Here are the examples Professor Craig (1997, p. 23) gives. Who are Michigan's two current senators? When did World War II begin and end? Who is the current US Secretary of State? Who was Joseph Stalin? These are factoids— informational tidbits that can be easily off-loaded and retrieved on a need-to-know basis. He does not include a single general principle such as "What conditions generally lead to instability in a country?" or "What are the biggest threats to democracy?"

Craig says he was dismayed to find out that his students were not embarrassed by their performance on his quiz, telling him (Craig, 1997, p. 23) that "they wouldn't need the information in their future jobs" and asking, "When is any of this stuff going to matter in my career?" On the basis of his short quiz and the students' subsequent defensive reaction to being told they had performed poorly on it, he then concludes that these students "see no need to understand why democracy needs to be preserved," closing with a dire warning: "...If our most promising young people have no appreciation for why democracy is worth preserving, how will they know when it is threatened?"

From the information presented in the essay, we side with the students. While a well-read, up-to-date person may have little trouble answering such specific and relatively trivial questions, college students are typically so busy with ample, challenging course work, jobs to pay for their education, and other college-related activities that most must virtually abandon public life during their college career. All of the questions cited have arguably little relevance to the students' immediate future, nor are their answers necessarily representative of the quality of their future citizenship, or even of their overall understanding of American history. While ignorance of dates and surnames is claimed by Craig (1997) to augur the demise of American democracy, we think it actually indicates college students' aversion to courses driven by obsolete views on what constitutes good instruction, and their rejection of educational practices that overvalue memorization and mindless learning.

Although we are not historians, we think the questions Craig (1997) presents as exemplars in no way comprise "basic historical and political facts" vital to American citizenship. We applaud the students who [we presume] were able to distinguish deep and powerful principles from trivial facts, to resist an unwarranted public intellectual flogging by Craig, and to argue that not all of one's knowledge of a domain must, of necessity, reside in one's memory. Detail recall alone doesn't prove that sound understanding exists. Unless detail is effectively integrated with a central knowledge structure, it merely adds to the data smog that obscures the subject. In the days before books and computers, wholesale memorization made more sense. But welcome to the dawn of the 21st century!

The strategies we present in this book aim to help students construct and polish useful central knowledge structures they need to master a new domain as well as to facilitate their meaningful acquisition of relevant detailed information. Of course, strategies can always be trivialized by instructors who haven't thought through important principles of their field and the purposes and goals of their instruction.

The days of teaching history as encapsulated by knowing the dates of wars and battles, or the surnames and titles of the powerful, are waning. The same is true for the teaching of biology as encapsulated by textbook terms, definitions, life cycle minutia, taxa names, and names of minor structures. Whenever you hear a colleague mourn that "Today's students don't even know what *Cnidarians* are (substitute any biological concept or name), first ask yourself if that missing knowledge might have been displaced by other knowledge of greater worth, or if such knowledge is actually a prerequisite for understanding very many of the important ideas in biology today. Such critical statements often say much more about the lamentor than they do about today's biology students.

Just as the person who is a "pack rat" soon can't find anything he's looking for among the burgeoning assemblage of his possessions, we suspect that, at particular points in one's biology learning career, too much trivial knowledge is a dangerous thing (or at least a cognitively burdensome one). The psychological study of human memory indicates that we forget most of what we don't use regularly and, over time, we consolidate our memories, so that only the most salient features remain.

Sometimes, focusing on too many details can actually obscure "big picture" views. Algebra provides a good counter example in that, by substituting letters in place of numerical detail, it allows us to see interesting mathematical patterns (e.g., the logic of triadic and higher relations) and also affords manipulatable formulae. Often these formulae reveal mathematical properties not otherwise discernible. Geometry uses indices in the form of letters of the alphabet affixed to diagrams to represent the spatial features asserted in its premises, and thus reveals generalizable spatial relationships as well. Once algebra is mastered, elaboration into numerical applications can take advantage of the foundational knowledge that preceded it.

To temper the foregoing assertions and to underscore that knowing biological principles doesn't make learning certain facts unnecessary, one of us (Fisher) is convinced that some factual knowledge provides an essential foundation from which abstract principles can be derived, and that without such knowledge, abstract principles have little meaning. For example, the general principle, "13. Life tends to optimize rather than maximize" (Hoagland & Dodson, 1995, p. 1), may make a lot of sense to biologists, but what does it mean to a novice? Such general principles require some concrete examples to illustrate the point, and it is difficult to know how many examples and how detailed they need to be to really get the general principle across.

Fisher also emphasizes that certain information is essential to work in a domain and simply must be learned, that a knowledge of facts readily available in the mind seems to account for a high percentage of expert performance (e.g., Gordon & Gill, 1989), and that acquiring factual knowledge can give a person attempting to master a new domain a good deal of intellectual satisfaction. The real issue is, what is the optimum balance between big ideas and detail? Are biology students today so deep in detail that they often miss the big picture?

THE "NEED TO KNOW" CRITERION AND
THE "DETAILS/PRINCIPLES" CONUNDRUM

Dennett (1996, p. 58) reminds us that, in the world of the spy, secret agents are supplied with only the information they need to know — and he says that nature seemingly employs the same criterion quite often. Consider the spores of bacteria, fungi, and ferns — they are marvels of minimalism! Ultimately, of course, more information is usually good. There is a growing body of evidence, for example, that 85–95% of an individual's ability to perform in a given domain is derived from the content knowledge about that domain that they have stored in and can readily retrieve from their memory (Gordon & Gill, 1989). *But there is a difference in acquiring detailed information for a purpose and working with it regularly to hone and polish its edges, and acquiring detailed information temporarily because an authority figure says that is what you have to do.*

We know of no research that demonstrates beyond doubt the optimum ratio of detailed knowledge to general principles in a given course, or the desired ratio of top-down to bottom-up instruction. However, we propose that common sense can go a long way toward finding the best ratio for a given class. Instructors need to ask themselves often, who are my students, what are their needs and interests, and what do I feel is important for them to know five years from now?

Students depend on their instructors to be selective in what they include and emphasize in a biology course. Any course is really just a small sample of the knowledge biologists have gained across the centuries. A sound cognitive infrastructure must be constructed that is able to bear the weight of additional and more detailed knowledge later, without collapsing. More information isn't always better, and everything isn't equally important for constructing such an infrastructure.

One problem is that biological research has become so specific and detailed, looking at minute details of molecular structures and interactions, for example, that it can be truly challenging for a researcher who is so focused to even perceive the big ideas. Yet this is the hallmark of a good teacher.

How is all of the foregoing information related to mapping biology knowledge? Answer: Learners need to integrate their new biology knowledge with existing knowledge, they need to organize their knowledge into coherent patterns, and they need to polish and refine their knowledge structures. Knowledge mapping can help at every step. Hoagland and Dodson's generalizations suggest how one builds and maps a cognitive infrastructure capable of bearing the weight of knowledge elaboration across a lifetime of biology learning. We now look to systems theorist James Grier Miller for additional relevant insights.

A SYSTEMS VIEW OF BIOLOGY: AN OVERVIEW OF JAMES GRIER MILLER'S WORK

Few biology educators or biologists are aware that a scholar named James Grier Miller developed a 1,100-page general living systems theory over the course of nearly 30 years (1978). While advances in systems theory and in biology have obviously occurred since then, we have found Miller's *opus magnum* to be not only insightful, but also useful for advancing our thinking about mapping biology knowledge structures. He proposes seven hierarchical levels of living systems, a hierarchy that has been extended today.

Miller was a graduate student of the philosopher, Alfred North Whitehead, a man who saw life as a multilevel system of systems. Whitehead holds that continuous dynamic process (or event) is the essence of all reality. To him, an *event* is "the organism or system at the present moment" and all events are interconnected (Miller, 1978, p. xiv). Whitehead's process-oriented thinking inspired Miller to undertake this massive project.

Miller was an outstanding scholar in his own right, the coiner of the term *behavioral sciences,* and a behavioral sciences professor, first at the University of Chicago and then at the University of Michigan and the University of Louisville. Over the years, he not only worked with Whitehead but also other renowned intellects such as Ralph Tyler, Enrico Fermi, and Leo Szilard. Miller's predilection for ordering and integrating knowledge, coupled with a wide-ranging curiosity and an interest in the living world, made him uniquely qualified to analyze the structure and process of living systems, helping to integrate our knowledge of the phenomena of life.

PRINCIPLES IDENTIFIED BY J. G. MILLER TO PROMOTE ORGANIZATION OF BIOLOGY KNOWLEDGE

In the second chapter we described a set of general principles in biology, as summarized by Hoagland and Dodson (1995). Here we review aspects of Miller's (1978) systems theory of biology. We see these viewpoints as providing valuable food for thought for biology teachers and learners. Biologists will be familiar with the many of these ideas — but how many convey such principles to their students?

Miller (1978) was likely unaware of the utility of concept maps and semantic networks for representing biology knowledge, but his work yields many insights that we think can be helpful to our readers in organizing and mapping biology knowledge. There is, of course, much more to be learned from Miller's masterpiece, but what follows here are the items that most captured our fancy. Perhaps they will inspire some of you to go to the source.

Miller on Living Systems

A system is a set of interacting physical units with relationships among them, all working together to perform a particular task or set of tasks (p. 16; see also Fig. 3.2). Basic to living systems theory are the concepts of space, time, matter, energy, and information (p. 9). In living systems, matter, energy and information are changing continuously across the dimensions of space and time. The structure of a system is a function of the arrangement of its materials, subsystems and components in three-dimensional space at a given moment of time (p. 22). Space can be either physical (as in a cell) or conceptual (as in the mind) (p. 10).

Living systems at all levels of organization are open systems comprised of subsystems which process inputs, throughputs, and outputs of various forms of matter, energy, and information (p. 1). That is, living systems can be described in terms of (a) input (substrates) and output (products); (b) flows through systems (for example, blood through the circulatory system, foodstuff through the digestive system, nitrogenous wastes through the urinary system); (c) steady states (such as those maintained by stable cells); and (d) feedbacks (regulatory elements) (p. 42). Living systems at all levels have 19 critical subsystems to carry out the major tasks of living organisms (p. 3). We have added some of the corresponding structures in humans as examples (shown in parentheses):

(a) reproducer of progeny (male and female reproductive systems);

(b) boundary between self and other (integumentary system);

(c) ingestor of nutrients (digestive system);

(d) distributor of material, energy, and information (blood and lymph systems);

(e) convertors of material, energy and information (muscles converting chemical energy to mechanical movement);

(f) producer of new organic matter (enzymes involved in synthesizing new molecules and macromolecules);

(g) matter-energy storage in organic molecules (glycogen in the liver; fatty deposits in cells);

(h) extruder of wastes (digestive system, urinary system, sweat glands);

(i) motor (mover) producing various types of movement at multiple levels including system, organism, tissue, cellular and intracellular (skeletal-muscular system);

(j) supporter of organismal structures (skeletal system);

(k) input transducer converting raw material to its own use (digestive enzymes to break down raw material; internal enzymes to assemble new molecules — see f);

(l) internal transducer, (throughput or t-transducer) converting material from one form to another in small steps (enzymes involved in all types of molecular conversions);

(m) channel and net (our interpretation is that channel refers to the brain and its ability to broadcast from working memory to all the modules of the brain (Baars, 1989), while net refers to the distributed nature of information in the brain (Rumelhart, McClelland & the PDP Research Group, 1987);

(n) decoder of information (brain decoding of sensory input);

(o) <u>associator</u> of events and experiences (associative memory);

(p) <u>memory</u> of past events, thoughts and fantasies (long term memory);

(q) <u>decider</u> of actions and attitudes (left hemisphere of the brain) (Jeannerod, 1985, p. 123);

(r) <u>encoder</u> of information (short term, intermediate and long term memory); and

(s) <u>output transducer</u>, sending materials and messages to other parts of the system or to other systems (signal molecules such as hormones and neuropeptides).

Systems can be differentiated along many dimensions. These include: (a) average size; (b) average duration; (c) amount of mobility; (d) degree of spatial cohesiveness; (e) density of unit distribution; (f) number of processes; (g) complexity of processes; (h) transferability of processes from one component to another; and (i) rate of growth (p. 28). The more two or more systems interact, the more they become alike in storing and processing common information" (p. 103). This is apparent today in the communication between the nervous and the immune system, which share common molecules for signaling cells (Clark, 1995).

These ideas are summarized in a concept map (Figure 3.2) (concept maps were introduced in Chapter 1).

Miller on Process & Change

Process is the change in matter-energy or information of a system over time (p. 23). Space-time coordinates are important for observing, measuring, and characterizing processes (Miller, 1978, p. 19). Use of buffer inventories between Process A and Process B eliminates the effect of Process A limiting Process B (p. 94). For example, if there is a good supply of tyrosine available in a cell, then the reactions that use tyrosine are not immediately dependent upon the reactions that produce tyrosine.

Time relationships within processes involve (a) containment in time (fertilization is followed by rapid mitoses to produce the blastula); (b) number in time (each chromosome is replicated once during interphase); (c) order in time (prophase follows interphase); (d) position in time (blastula formation occurs in the early developing embryo); (e) direction in time (mitosis always progresses forward, never in reverse – see g); (f) duration (a given adult cell may require 14 hours to complete mitosis); and (g) pattern in time (mitosis includes the successive stages interphase-prophase-metaphase-anaphase-telophase and is followed by cytokinesis) (p. 281).

Boundaries such as the plasma membrane or mitochondrial membrane serve as barriers to matter-energy and/or information flow. They prevent certain matter-energy and/or information from entering or leaving and maintain a steady-state differential between the interior and the exterior of the system (p. 56).

Many aspects of systems' historical processes depend upon the balance at any period between anabolism and catabolism, growth and decay. In youth anabolism is dominant, in maturity the processes are in balance, and in later years catabolism usually predominates, producing senescence and decay (p. 82).

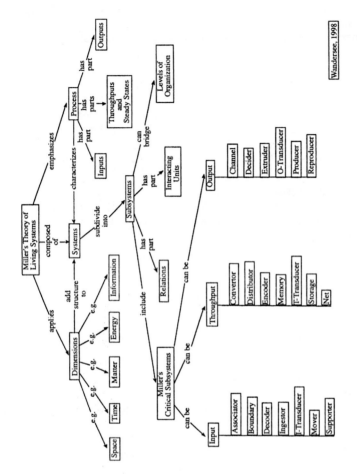

Wandersee, 1998

Figure 3.2. Concept map of Miller's theory of living systems.

Miller on the Continuity of Life

All nature is a continuum (p. 1025). Reproduction is critical to the continuity of the species or system (p. 1). Evolution is like learning, but the species, not the individual, does the learning (p. 77). Levels of living systems differ as to time periods of origin (p. 1035).

Miller on Information

Living systems divide the intensities of information inputs into about seven categories, plus or minus two (p. 95; first proposed by George Miller, 1956). The less decoding and encoding an information channel requires, the more it will be used (p. 97). The cost of keeping large masses of information is high (p. 640). The longer information is stored in memory, the harder it is to recall (especially if it isn't used regularly) and the more likely it is to be incorrect (p. 99). Meaning is the significance of information to a system that processes it (p. 11). Development of a rigorous and objective method for quantifying meaning would be a major contribution to the science of living systems (p. 12).

Miller on the Importance of Theory

There is a need for integrative theory to encompass the huge volume of research findings and close the gaps in our knowledge of living systems (p. 5). Science must collectively fill in the conceptual gaps in order to achieve such knowledge integration (p. 1051). In fact, scientific journals are like catalogs of spare parts for a machine that is never built – that is, many of the published parts are never assembled into a coherent theory.

The parts can tell us about one another. For example, if two parts are interrelated either quantitatively or qualitatively, knowledge of the state of one must yield some information about the state of the other (p. 13). Cross-level generalizations will prove fruitful in the study of living systems and can have great conceptual significance (p.26).

Miller on Biology Knowledge

We are used to seeing the world as a collection of concrete objects (noun things) in space-time, and these objects naturally draw our attention" (p. 21). Rank ordering of the types of things always varies, depending on the classification variable we select (p. 25).

Most biological hierarchies are described in spatial terms (Miller, *pp 17 to 19*; first proposed by Simon, 1962, p. 469). The relations among components of living systems are not imaginary; they are inherent in the totality of the system (p. 1051). That is, while the names of relations may be arbitrary and will vary with both the language spoken and with individual preference (e.g., *has part* or *has component)*, the relations themselves are inherent in the structures and interactions.

Miller on Scientists

We tend to focus our attention on a single plane of analysis and ignore the rest (p. 1050). Observation always is shaped by the purposes and characteristics of the observer as particular sets are chosen for study from the infinite number of units and relationships available (p. 16). Specialization is a way scientists protect themselves from information overload (p. 1050). Scientists are laymen to all scientists except their fellow specialists, but there is an unspoken agreement of "uninformed mutual acceptance" (p. 1050) because of the common or shared values involved in examining and describing things and in generating new knowledge.

Miller on Biology Teaching: A Caveat to Ensure Understanding

In biology teaching, every discussion should begin with an identification of the hierarchical level of reference, and the discourse should not change to another level without a specific statement that this is occurring (p. 25).

AN EPILOGUE TO OUR OVERVIEW OF MILLER'S WORK

A systems view of life is highly compatible with the approaches to mapping biology knowledge that are presented in this book. The propositions extracted from Miller's (1978) opus were chosen for their relevance to biology knowledge organization and biology mapping efforts, as well as to the higher level of understanding afforded by general principles.

For example, his analysis of time relationships within processes are helpful in describing the temporal dimensions of a biological process, reminding us of the many ways that time-based descriptors can be used to build a knowledge structure that captures the important features of our own biological understanding.

He advises us to include cross-level generalizations in our knowledge structures – something often overlooked and which, according to Miller (1978), can have a potentially large payoff in understanding. His comment about the nature of scientific journals reminds us that the goal and measure of progress in science is theory building, not publishing insular research papers.

His observation that the more two or more systems interact, the more they become alike in storing and processing common information is anticipatory of recent findings about the high level of interactions between the nervous and immune systems, their abilities to read one another's signals, and to produce some identical signal molecules (Clark, 1995).

SUMMARY

We began this section by noting Whitehead's view of life, that continuous dynamic process (or event) is the essence of all reality. Process does go on nonstop and this view often escapes the novice, but it is also important to recognize that all of biology arises from the intimate relation between structure and function.

Furthermore, a small change in structure can produce a profound change in function. Consider the deadly genetic diseases produced by a *single base change* in the 3.2 billion nucleotide base pairs that comprise our DNA, as in phenylketonuria and thalessemia. Consider that whether a child will be male or female is determined by small differences in side groups on the steroid core that comprises testosterone and the estrogens.

This entire chapter is part of our prelude to mapping and was written with the dual aims of a) introducing a valuable resource for biology knowledge organization and b) providing some scaffolding with which one can select, organize, integrate, and refine biology knowledge to create an informed, customized, and powerful personal knowledge structure. The use of ABO blood types by the Japanese to predict future personalities is illustrative of the overall future orientation of our cognitive framework — an orientation that can help teachers design better courses by asking themselves, What do I want my students to know five or ten years from now? The need-to-know principle can be used like Occam's razor, to preserve relevant detail and cut away the clutter. Our long term goal is to help biology students learn mindfully and meaningfully (see Chapter 5), but to achieve that, it is clear that many biology teachers need to rethink the ways in which they organize their courses. Trivial pursuit does not belong in the classroom. Categorizing ideas and organizing them into hierarchies facilitates learning biology, and this can be achieved by using a systems approach to analyze living systems. The view from the top of a mountain (hierarchy) can help a hiker find his/her way through the terrain. Detailed knowledge is usually required for generating general principles, but it is the general principles that are best remembered.

And while many biology courses need rethinking, in our opinion, biology textbooks need it even more. The prevailing formula for science texts inhibits comprehension and meaningful learning to a large degree. With few exceptions, science texts are well-organized compendia of currently important facts, with few clues as to the connections among them or why they are considered important.

Our minds are facile story apprehenders that require context and flow to make sense of events, but in biology texts the chronology and relatedness of the story has been stripped from the factoids and relegated to the sidebars. The naked details and occasional generalizations that remain do little to promote the kind of learning we would all like to see in our biology students, namely, meaningful and mindful learning.

Our minds are also spatial navigators, linking events to loci in our environment. In a biology course, students like to know where they are now, where they are going, and where they've already been. When he was writing his books, the Nobel laureate author, Ernest Hemmingway, would only stop working when he knew exactly what he was going to write next. That way, it was easier for him to get started again on the following day. When we are in a large and unfamiliar shopping mall, the helpful store directory kiosks typically have a map of the shopping center with a benchmark labeled "You are here." Meaning is always referential. No benchmark...no coherent understanding. The same is true for biology learning and this is a far-from-trivial point. Miller knew this and his penchant for underlying benchmark principles helps us to deal with the complexity of life in the biosphere.

Making connections between our students and the world they live in is important. This only occurs when teachers possess a "big picture" view of biology. The "big picture" view requires a well-organized and well-integrated knowledge base supported by an appropriate level of detail.

KATHLEEN M. FISHER & DAVID E. MOODY

CHAPTER 4

Student Misconceptions in Biology

Achieving Understanding

A number of young children were given a log and asked, "Where does the weight of this dry log come from?" They responded, "from the sun, water, the soil, the seed. . ." Harvard and MIT graduates were then asked the same question. Their answers were largely the same as those offered by the youngsters. The university graduates were then asked, "What would you say to someone who said to you that the weight of the tree comes mostly from carbon dioxide in the air?" Among their replies were such comments as: "Really! I would wonder about that. I would wonder how that's possible." "I would disagree because this same volume of air would weigh much less unless it was highly compressed." "I'd say obviously, carbon dioxide is intimately involved in photosynthesis. I'd say carbon is not much of a building block from what I know of biochemistry." "I'd say that's very disturbing and I wonder how that could happen."

A middle school student, Jon, is given lessons about photosynthesis. After the lessons he knows the formula for photosynthesis by heart and is able to write it on the board. When asked what is in the dry log, he estimates that it contains about "70–75% water, and 25–30% other stuff, including bark and minerals from the soil." The interviewer asks if any of the carbon dioxide that goes into the leaves stays in the tree. Jon says, "Uh, maybe a little bit but not too much." His logic is that if oxygen and carbon dioxide had weight, we couldn't breathe; and if these gases have no weight, then carbon dioxide cannot possibly contribute to the weight of a tree.

The interviewer gives Jon a block of dry ice to hold with an asbestos glove and tells Jon that this is frozen carbon dioxide. Jon is very surprised as he realizes the block has weight. The interviewer asks what Jon thinks about the weight of CO_2 now. Jon concludes that solid carbon dioxide might have weight but gaseous CO_2 does not.

Excerpts from Schnepps, M. (1997b).

THE NATURE OF MISCONCEPTIONS

Among the most pervasive features of the terrain uncovered by the cognitive paradigm in science education is the presence among many students (in fact, probably among all people) of active misconceptions about natural events. The "blank slate" model of the mind postulated by Locke (Locke, 1891, 1996) encouraged the easy assumption among educators that students receive instruction as if they were empty vessels, devoid of any ideas of their own.

In fact, we now know that students come to the classroom brimming with ideas about a great many issues and events in the natural world. People are constantly building mental models to make sense of the world around them. Unfortunately, a substantial number of these models are erroneous from the scientific point of view. For example, the common and persistent misconception that carbon dioxide cannot contribute to the weight of a tree has been extensively studied (Haslam & Treagust, 1987; Wandersee, 1984; Anderson, Sheldon & Dubay, 1990; Gravett & Swart, 1997). The identification and description of such naive ideas represents a major stream of activity in science education research.

Within the research community, a profusion of names has been suggested to refer to such conceptions, reflecting the dynamic and unsettled nature of the field. Many investigators prefer the designation 'alternative conceptions,' since it is value-neutral and demonstrates respect for student ideas. Other proposed names range from the simple — "naive ideas," "prescientific ideas," "preconceptions," and "conceptual primitives," to the complex — "limited or inappropriate propositional hierarchies" or LIPHS (Wandersee, Mintzes, and Novak, 1994). The present chapter prefers to adopt an eclectic approach in which varying terms are employed according to their nuances and context. The primary term employed here, however, remains "misconception," selected to underscore the cognitive transformation required in order to achieve the scientific view.

The discovery of the fertile field of students' conceptions suggests a modification to Ausubel's dictum (1963, 1968), "Ascertain what the student knows, and teach accordingly." With the recognition that what the student "knows" consists in part of ideas that conflict with scientific beliefs, Ausubel's admonition might more appropriately be stated, "Ascertain what the student misunderstands, and teach accordingly." This injunction, however, turns out to be more difficult to put into practice than it may at first appear to be.

The purpose of this chapter is to elucidate the nature of preconceptions, and to suggest ways in which mapping devices such as circle diagrams, concept maps, mind maps, and SemNet can be employed as a kind of bridge to enable students to make the transition to the scientific view.

Wandersee, Mintzes, and Novak (1994) have reviewed more than 3,000 studies of misconceptions in science. They distilled eight propositions that represent the consensus of investigators. These are summarized (in a different order from the original) and elaborated upon below.

First, learners come to formal science instruction with a diverse set of alternative conceptions concerning natural objects and events. To a large extent, these alternative conceptions are widely shared, often held by 20% or more of a given student population. Science teachers are largely unaware of the existence of these ideas in students' minds.

Second, the alternative conceptions that learners bring to formal science instruction cut across age, ability, gender, and cultural boundaries. Many similar conceptions are found in students and in the general population worldwide.

Third, alternative conceptions often parallel explanations of natural phenomena offered by previous generations of scientists and philosophers. The fact that these naive conceptions are widely shared across both space and time is a tribute to their sensibility. They are logical conclusions drawn from limited data. Further, they underscore the point that many scientific ideas are counterintuitive. Scientific understandings represent hard-won insights into the workings of the world.

Fourth, alternative conceptions have their origins in a diverse set of personal experiences including direct observation and perception, peer culture and language, as well as in teachers' explanations and instructional materials. They are a product of active sense-making (see also Chapter 5).

Fifth, for the reasons described below, alternative conceptions are tenacious and resistant to extinction, especially by conventional teaching strategies. Where such conceptual conflicts are concerned, students often require compelling evidence – they truly need to be convinced. Simply being told is not sufficient reason for them to dismantle their well-established belief systems. The students' own ideas are so well established and so satisfying to them that they tend to be reluctant to replace them with scientific ideas. The scientific ideas may be rejected because they seem foreign, silly or unbelievable, as well as because of the emotional attachment students have to their own ideas. In other cases, the scientific ideas may be altered or misinterpreted so they can appear to be consistent with the student's ideas.

Sixth, to complicate matters further, teachers often subscribe to the same alternative conceptions as their students. As noted above, nonscientific conceptions are not limited to students; they are as natural to human beings as breathing. We all have them. They occur because most people, scientists included, do not employ the scientific method in their everyday efforts to make sense of the world. Nor do most people have access to the accumulated wisdom of every field. Humans simply draw the best conclusions they can on the basis of what is usually their limited knowledge.

Seventh, learners' prior knowledge interacts in profound ways with knowledge presented in formal instruction, resulting in a diverse set of unintended learning outcomes. Many teachers assume that "I told them, they heard me, therefore they know it." This, in fact, may be the most widespread misconception in education.

Eighth, instructional approaches that facilitate conceptual change are usually essential for replacing a resistant misconception with a scientific idea. Such approaches are generally difficult to discover and time-consuming to implement. But effective conceptual change strategies are at the heart of inquiry-based science teaching and constructivist learning. They are necessary if the American public is going to acquire even a modest degree of sophistication in scientific thought (see, for example, McComas, 1997).

In summary, alternative conceptions are not idiosyncratic or peculiar to individuals or groups of individuals. On the contrary, they are shared across age, gender, and culture, they appear regularly in the history of science, and they occur in the cognitive structures of many adults. Preconceptions are not arbitrary or random explanations for events, but rather represent a pattern of understanding that is plausible to the learner who is attempting to make sense of the world with limited knowledge.

The review by Wandersee, Mintzes, and Novak (1994) enables us to see the fundamental characteristics that are shared by misconceptions. Other resources which summarize research on misconceptions include Helms and Novak (1983), Champagne, Gunstone, and Klopfer, (1985), Novak (1987, 1993), and Pfundt & Duit (1994).

One positive aspect of misconception research is the attention it has brought regarding the absolute necessity for teachers and researchers to be well-grounded in both content knowledge and pedagogical content knowledge. That is, to be good at what they do, researchers and teachers must know at a deep level both the content being taught and the specific strategies useful for teaching that topic, known as pedagogical content knowledge or PKG.

It is also important to recognize that preconceptions are not exclusively obstacles to learning. Since preconceptions often have some predictive power in certain practical situations, Clement (1982a) suggests that they be thought of as zero-order models which can be modified with appropriate instructional strategies.

The fact that both useful prior knowledge and misconceptions exist in abundance is a reasonable and straightforward consequence of personal knowledge construction and strong verification of constructivist learning theory (Pope, 1982; West and Pines, 1985; Clement, 1982a; Collins & Gentner, 1982; von Glasersfeld, 1987; Fisher, 1991; Gunstone, 1994). Students are actively engaged in making sense of the world around them long before they arrive at the classroom door. If many of their ideas about natural processes are naive and contradictory to scientific ideas, that is merely indicative of the fact that the findings of science are often counterintuitive or otherwise not obvious. Indeed, if everything in nature were just as it first appears, science would hardly be necessary at all.

SOME EXAMPLES OF COMMON MISCONCEPTIONS IN BIOLOGY

Few biology faculty are aware of the obstacles their students face in trying to come to terms with even simple biology ideas. The vignette at the beginning of this chapter describes a well-studied misconception that is highly resistant to change – namely, the belief by many people that an invisible gas, carbon dioxide, cannot possibly contribute carbon to growing plants for making sugars, starches, and cellulose. The problem is that a great many people believe that gases have no weight because we cannot feel the air around us. This primitive belief interferes with the learning of many science ideas in addition to photosynthesis, such as changes of state, conservation of matter, and so on.

As another example, one of us (Fisher) finds that up to 20–25% of her college seniors every semester do not understand what makes up the bubbles in boiling water. They claim that the bubbles contain oxygen and hydrogen, or air, or sometimes a vacuum. Convincing the students that the bubbles contain water vapor is no easy task – again, telling is not enough. This conception comes from a lack of understanding of changes of state, conservation of matter, and also from the common belief that you can see water vapor, but when water evaporates it turns into an invisible gas and therefore is not water vapor.

A third example is the difficulty involved in understanding what it means to be alive (Stepans, 1985; Carey, 1987; Tamir, Gal-Choppin, & Nussinovitz, 1981; Brumby, 1982). Young children often think that plants are not living because they are not mobile, and many older students assume that such life forms as seeds are not alive.

DISCOVERING MISCONCEPTIONS

People who first read or hear about misconceptions imagine that they must come tumbling out of students' mouths in every classroom. If this were the case, students' naive conceptions would have been discovered long ago. On the contrary, several factors conspire to keep teachers from ever knowing what students are really thinking. First, students generally have implicit rather than explicit knowledge, meaning that they are not quite aware themselves what they are thinking or what assumptions they are making. Second, students are not encouraged to say what they are thinking in traditional classrooms, so that even when their knowledge is explicit, students learn to keep it to themselves. Third, the opportunities for students to express themselves in nonverbal ways in today's classrooms are severely limited, since so much testing is now multiple choice and short answer. And fourth, teacher-designed multiple-choice tests offer what the teachers consider to be valid distracters, not what the students think. Most naive conceptions are so far removed from the scientific view that it simply doesn't occur to most teachers to include such ideas in their test items.

Identifying and characterizing naive conceptions generally entails considerable effort. One of the most frequently used techniques for eliciting students' ideas is the clinical interview (Pines, Novak, Posner, & VanKirk, 1978; Osborne & Gilbert, 1980; Ericsson & Simon, 1984). Two other frequently used methods described in more detail below are concept maps and multiple choice tests which incorporate common misconceptions as item distracters. Other approaches have used sorting and word association tasks (Champagne, Gunstone & Klopfer, 1985) and computer simulations (Nachmias, Stavy & Avrams, 1990).

DISTINGUISHING MISCONCEPTIONS FROM OTHER ERRORS

There are many different kinds of cognitive errors such as a slip of the tongue (Brown & O'Neill, 1966), action slips (Norman, 1981), and information processing errors (Fisher & Lipson, 1985). These types of errors are usually easily corrected. As noted above, naive conceptions are set apart from these errors in that they are shared by a significant fraction of students; they are surprisingly resistant to being taught away (especially with traditional, didactic teaching methods); and they often appear in similar frequencies in classrooms around the world.

Resistance to change is their most pronounced feature and the one that is most troublesome to teachers. In many cases naive conceptions are so deeply embedded in an individual's conceptual ecology that contradictory information either bounces off or is modified to fit the preexisting theory. The cognitive upheaval that is necessarily associated with dismantling a deep-seated belief is avoided by rejection or modification (often subconscious) of the new contradictory idea. It has become clear that students can go through course after course, even through entire majors, yet remain impervious to key scientific ideas being presented to them.

CONFLICTING BELIEF SYSTEMS: A SPECIAL PROBLEM IN BIOLOGY

> The Darwinian revolution has been called, for good reasons, the greatest of all scientific revolutions. It represented not merely the replacement of one scientific theory ("immutable species") by a new one, but it demanded a complete rethinking of man's concept of the world and of himself; more specifically, it demanded the rejection of some of the most widely held and most cherished beliefs of western man. ... In contrast to the revolutions in the physical sciences (Copernicus, Newton, Einstein, Heisenberg), the Darwinian revolution raised profound questions concerning man's ethics and deepest beliefs. (Mayr, 1982, p. 501)

People in Darwin's time thought of themselves as special beings, created by God in the form of Adam and Eve, and entirely separate from the kingdom of living things. The tenets of neo-Darwinian evolution eliminate that special position of humans in the universe. Today we know that 97% of the 3.2 billion base pairs in human DNA are arranged in a sequence that is identical to that in chimpanzees, and we know thousands of other facts that clearly indicate not only relatedness among all living things, but the degree of relatedness between any two species. In spite of the overwhelming evidence, however, the unity of all living things remains difficult for many people to accept.

According to Mayr (1982), an expert on Darwin and evolution, there are two key reasons why the concept of natural selection was so alien to the western mind prior to the nineteenth century. One was the pervasive prominence of essentialism in western thought. Essentialism, handed down from Plato, is the assumption that all members of a species are virtually identical: They share the same essences and are immutable. As Mayr says, "It is quite impossible to develop evolutionary theory on the foundation of essentialism. . . . Since they [species] lack variation, they cannot evolve or bud off incipient species" (p. 407).

The second was the equally pervasive idea that the Creator's design automatically conferred perfection. Given this assumption, there would be no room for improvement of a species or adaptation to new environmental conditions via natural selection. Mayr identifies a variety of other assumptions in western thought that are also in conflict with evolution.

The difficulties created by conflicting cultural beliefs are compounded by the inherent difficulty of the subject. The theory of evolution is supported by data from many different fields, most of which are not well understood by beginning biology students. To understand neo-Darwinian evolution, one has to master many interrelated, abstract concepts and the complex relations among them. A reasonably high level of motivation is required to achieve understanding of a domain as complex and subtle as evolution. Individuals who lack that motivation find it easy to dismiss the topic altogether.

ADDRESSING THE CULTURAL CONFLICTS

The bottom line is that biology teachers face enormous challenges when teaching evolution. How can a teacher address the conscious conflict that occurs when students feel that evolutionary theory threatens their religious convictions? One way to defuse the situation is to explicitly assure students that they have the right to their own ethical and religious beliefs. At the same time, as biology students enrolled in a biology course, they have a responsibility to learn biology. These two conditions are not mutually exclusive.

An analogy may help. The authors don't believe in Santa Claus and yet they know a great deal about the gentleman, such as where he lives, what he does, how he dresses, who he works with, how he gets around, what he looks like, and to whom he is married. This illustrates that it is possible to learn about an idea without believing in it.

When students are assured that their personal beliefs are respected and not under attack, but that they are nonetheless expected to master the ideas of evolution as an intellectual endeavor, much of their resistance to learning is reduced. To create and maintain a "safe" classroom, an instructor can show respect for divergent beliefs in many ways, as by asking questions in the form of "Why do biologists think that...?" or "What evidence led biologists to believe...?" This approach helps students to maintain an intellectual separation between their personal beliefs and the scientific ideas.

Strike and Posner (1985) suggest that belief in an idea is the final and necessary step to complete the conceptual change process. Disbelievers of evolution may not take this final step. But if students can achieve a reasonable level of understanding of evolutionary theory, I think most teachers would be satisfied.

Jackson, Doster, Meadows & Wood (1995) comment on a tendency among science professionals to view or treat orthodox Christian students in an unconscionable manner — to disrespect their intellect or belittle their motivations, to offer judgments based on stereotypes and prejudices, and to ignore the threat that evolution can pose to students' self-esteem. A respectful and tolerant attitude on the part of the instructor can go a long way to reducing tensions and promoting learning.

Addressing the second problem, students' implicit alternative conceptions and background assumptions, is even more difficult. Inquiry-based active learning, incorporation of many actual examples from and studies in evolution, and conceptual change strategies are all in order. Time is also an important factor, and in most biology classes, not enough time is allowed. The greater the cognitive distance to be traversed, the more time is required. Consider the amount of time that Darwin spent getting comfortable with and convincing himself of his radical ideas.

One of the issues in biology that first caught the attention of investigators was student understanding of the process of natural selection. In 1979, Margaret Brumby demonstrated a rather remarkable finding: First-year University students in England, most of whom had passed advanced ("A-level") courses in biology or zoology in secondary school, were often unable to give a correct account of this fundamental process. Instead, students tended to employ an alternative explanation for evolutionary events, one that bore little resemblance to the Darwinian concept. Brumby's results have been replicated by subsequent investigators working with students at various age levels in many different countries (e.g., Bishop and Anderson, 1990; Demastes, Settlage & Good, 1995). Aleixandre (1996) has identified a variety of alternative conceptions related to natural selection (Table 4.1). We reproduce it here with permission because it illustrates the complexity of the problem in bringing about conceptual change.

Aleixandre's list includes one of the most persistent preconceptions, that environmental pressure *causes* variation within a population in the manner originally postulated by Lamarck. Students with this view believe that ducks developed webbed feet and giraffes developed long necks because they *needed* them, not because a spurious mutation introduced these features into the populations and they proved to be advantageous.

Table 4.1. Facets of the Neo-Darwinian (D) and common alternative views, which in many cases are Lamarkian (L) in nature, of evolution. Adapted from Aleixandre (1996). Items have been reworded, modified and reorganized, but the original numbering is retained. There is a controversy today about antibiotic and insecticide resistance (D10/L10), such that the positions described below are not so clear-cut.

Neo-Darwinian View	Common Alternative View
D1. Individuals within a species vary in significant ways.	L1. Individuals within a species are all essentially alike.
D3. More offspring are produced by parents than can survive.	L3. Most offspring can be expected to survive.
D5. Individuals with an adaptation favored by the current environment will generally contribute more offspring to the next generation.	L5. Individuals respond to a changed environment by becoming slightly better adapted to that environment with each successive generation.
Ideas About Genetics	
D2. Mutations are chance events.	L2. Mutations occur in response to needs.
D8. Inherited traits are those that are specified by alleles contained in the gametes of the parents and in the cells of the individual.	L8. Inherited traits include traits acquired by the parents (such as physical fitness or short tails) that are passed on to offspring.
D10. Antibiotic or insecticide resistance arises in individuals through random mutation and in populations through multiplication of the resistant type.	L10. Antibiotic or insecticide resistance is acquired by individuals in response to exposure to antibiotic or insecticide. (but see figure caption)

Table 4.1. (continued) Facets of the Neo-Darwinian (D) and common alternative views, which in many cases are Lamarkian (L) in nature, of evolution.

Ten Ideas About Population	
D4. Competition occurs among members of a species for space, food, etc., which usually resolves in favor of those best suited to the environment.	L4. Competition occurs among members of the same or different species and usually takes the form of physical combat; the strongest individuals win.
D6. Populations change as the proportion of individuals in the population with a given trait changes.	L6. Populations change through gradual changes in a trait which occur within each family over successive generations.
D7. A change in the environment may result in an increased proportion of a trait in a population by favoring individuals having that trait.	L7. A change in the environment induces changes in individuals so they will be better adapted to that environment.
D12. A species can become extinct even if it plays an essential role in the ecosystem; in fact, 99% of all organisms that ever lived are now extinct.	L12. A species that is essential in an ecosystem cannot become extinct.
D14. Proportions of traits among offspring are probabilistic and therefore will vary somewhat from expectations.	L14. Proportions of traits among offspring are determined and will occur regularly.
Ideas About Individuals	
D9. Organs often exist even though they are no longer used; such organs are lost from the species only by chance mutations.	L9. Organs that are no longer used atrophy and gradually disappear from the species.
D10. Adaptive traits arise by chance genetic processes and cannot be willed into existence.	L10. Adaptive traits arise in response to environmental pressures and reflect the will to survive.
D13. *Homo sapiens* is subject to the same rules as other species – that is, humans can become extinct, and species useful to humans can become extinct.	L13. *Homo sapiens* is exempt from the rules of nature – that is, humans will not become extinct nor will species useful to humans become extinct.
Ideas About Time	
D15. The time scale in evolution is measured in eons.	L15. The time scale in evolution is measured in a few generations.

However, we believe there are semantic problems in the "need" questions frequently used in science education research. At one level, organisms do need certain traits to survive in certain environments. Dolphins need a sleek body to swim in the water. Desert plants needs ways of conserving water. In our clinical interviews with college students, we found that students who believe change occurs through random mutation and then may be favored by natural selection are as likely to refer to the "need for certain traits" as students who think the changes themselves result from environmental pressure. Beliefs about evolution are, in fact, very difficult to assess.

Students may occasionally hold a "pure" alternative view involving all fifteen alternative conceptions enumerated above, but more often they have a mixture of neo-Darwinian and alternative facets. Different students have different mixtures of ideas, and the distinctions are often subtle.

Further, the phenomena themselves are not always simple and straightforward. For example, biologists now believe that at low doses of insecticides, some insects can acquire resistance. This makes the alternative conception, "L10, Adaptive traits arise in response to environmental pressures and reflect the will to survive," at least partially correct some of the time. Molecular biology mechanisms are elegant, relatively simple, and ubiquitous in living things. Occam's razor prevails. Evolutionary mechanisms are also largely shared, but evolutionary outcomes are varied indeed. Consider the cheetah and the antelope it chases. Each has evolved the capacity for awesome speed, but they have developed their abilities by two entirely different methods. This in itself is consistent with evolutionary theory: Given the strong role of random chance, could we expect anything different? As Gould (1994) says, it is impossible to predict the twists and turns of evolution. Gould's words are worth repeating: ". . . [evolutionary] theory can predict certain general aspects of life's geologic pattern. But the actual pathway is strongly *undetermined* by our general theory of life's evolution. . . Webs and chains of historical events are so intricate, so imbued with random and chaotic elements, so unrepeatable in encompassing such a multitude of unique (and uniquely interacting) objects, that standard models of prediction and replication do not apply." (Gould, 1994, p. 85).

Maybe in another hundred years some of the old preconceptions will have faded away. Maybe the growing bonds between the natural sciences and the social sciences will create world views that are better able to facilitate science learning, as suggested by Cobern (1996). Maybe some missing pieces will be discovered that will make evolutionary theory simpler. In the meantime, teaching and learning evolution will remain an extremely challenging task. For more information on evolution education see Good, Trowbridge, Demastes, Wandersee, Hafner, & Cummins (1992).

RESOLVING MISCONCEPTIONS IN FLASHES OF INSIGHT

Overcoming a misconception can be equivalent to a flash of insight (Moody, 1993). Consider the following story, for example. Hieron II, king of Syracuse, posed a puzzle for Archimedes, his chief mathematician, inventor, and sage. A neighboring king had given Hieron a crown that was supposed to consist entirely of gold. Hieron was somewhat doubtful of this claim; he suspected that the gold had been alloyed with a significant portion of silver. He asked Archimedes how he could determine (without actually melting down the crown) whether or not it was in fact composed entirely of gold.

For days Archimedes labored over this problem, but without success. Yet neither was he able to put the puzzle aside, for something about it suggested to him that the solution was near at hand. After studying the issue from every angle, however, nothing approaching a satisfactory answer occurred to him. Perhaps, he concluded reluctantly, it would be better to take a rest from this intractable problem, and so he asked his servant to draw his daily bath. Archimedes undressed and began to settle into the warm water. As he did so, the water level rose and overflowed the tub. Suddenly Archimedes leapt up from his bath and ran out into the street. "Eureka!" he shouted in naked exuberance. "I have found the solution!" What occurred to Archimedes is that equivalent weights of different substances (such as gold and silver) occupy different volumes. By closely observing what volume of water the crown displaced, therefore, it would be possible to determine whether it was made entirely of gold.

THE COGNITIVE STRUCTURE OF MISCONCEPTIONS

The "Eureka" moment of Archimedes may be considered a model of the occasion in which a constellation of elements suddenly and instantaneously arrange themselves in a new configuration – a true moment of insight. It illustrates the sudden coming together of a new perspective, a fresh new view of the problem which happens also to contain its potential solution. In the case of Archimedes and in fact, whatever the issue at hand, the success of an insightful event depends upon a rearrangement of cognitive elements rather than the mere addition or deletion of a concept.

Students sometimes experience insight as well, as demonstrated by Demastes, Good, and Peebles (1996). They investigated the patterns of conceptual restructuring by students within the theoretical framework of biological evolution. Demastes et al. found four patterns of conceptual change: cascade, wholesale, incremental, and dual. The first two patterns appear to be consistent with moments of insight.

These examples illustrate the cognitive structure of conceptions and the ways in which they change. Moody (1993) has proposed that the extent of the differences between a naive conception and its corresponding scientific conception can be described as *"cognitive distance."* Where the transition involves an extensive reorganization of relationships among conceptual elements, the cognitive distance is great; where the reorganization is less extensive, the cognitive distance that must be traveled to correct it is smaller.

We suggest that with misconceptions, the cognitive distance would generally be expected to correlate with the degree of resistance to change, although other factors may be involved as well, especially the strength of various elements of the naive conception in the individual's worldview.

In a section above, we drew a clear distinction between alternative ideas and simple errors or mistakes. For the purposes of illustrating cognitive distance, however, we will consider errors and misconceptions as points on a continuum, rather than as sharply distinct cognitive phenomena. Consider, for example, the following set of erroneous propositions:

1. The Earth is the fourth planet from the sun, rather than third.

2. The Earth is round like a pancake, rather than round like a sphere.

3. The seasons are caused by variations in the Earth's distance from the sun, rather than by the tilt in the Earth's axis of rotation.

If a student believed the first proposition to be true, the kind of shift required to correct the error would amount to no more than an alteration of one or two links in the student's cognition. Such mistakes do not qualify as misconceptions, but they could be considered points at one pole on a continuum of increasing complexity.

If a student believed the second proposition to be true (and this misconception is well-documented in the research literature), the kind of shift required to correct the naive notion would amount to a more significant alteration in the student's cognition. The notion that the Earth is round like a pancake is evidently appealing to young children because it reconciles the intuitive sense that the Earth is flat with the received idea that the Earth is "round." In order to understand the actual shape of the Earth requires at least two significant changes in cognition. First, it requires an expansion in the meaning of "round." In addition, it requires some appreciation of the size of the Earth, in order to understand that the planet could appear to be flat while in fact being spherical. The complexity of the cognitive change required in this case qualifies the naive notion as a misconception. Among all misconceptions, however, this one is relatively simple, which no doubt accounts for the fact that it is successfully overcome by all but a small fraction of the population.

If a student believed the third proposition to be true (and this misconception is also well-documented), the kind of shift required to correct it would entail alteration in a whole series of links in cognition. In order to understand the origin of the seasons, students must come to understand the complex relationships among many ideas, including the axis of the Earth's daily rotation, the tilt of the Earth's axis relative to the plane of its orbit around the sun, the resultant changes in the angle of incidence of the sun's rays upon the Earth, and the ways in which changes in the angle of incidence affect the amount of heat that is absorbed. The cognitive shift required to achieve the scientific view is therefore rather complex, which probably accounts for the fact that the naive notion remains well-entrenched among college graduates.

We may consider these three erroneous propositions on a continuum of increasing complexity. Within this continuum, a minimum level of complexity would be required in order to mark the transition from a simple error to a misconception. This notion of complexity, in turn, may give rise to the construct of "cognitive distance", which may be defined as the degree or extent of change required in order to achieve the scientific view.

Another important dimension in thinking about misconceptions is whether a naive conception stands alone or is derived from a deeper underlying misconception, as illustrated by the way in which the assumption that air has no weight interferes with learning photosynthesis. Addressing the common underlying misconception could conceivably resolve many learning difficulties at once, and so such information is potentially quite valuable, especially for designing the K–12 curriculum (see the curriculum discussion below).

DISPELLING MISCONCEPTIONS WITH
CONCEPTUAL CHANGE STRATEGIES

As we have seen, the constructivist theory that predicts misconceptions carries a corollary: the understandings that the learner develops in the effort to make sense of experience are not easily put aside (Hewson & Hewson, 1988; Posner, Strike, Hewson, & Gerzog, 1982; Strike & Posner, 1985). Instead, the ideas the learner constructs are robust in their ability to withstand correction. To change a misconception is often like remodeling a house. It requires extensive dismantling and rebuilding.

Clement (1982b, 1987), a physics educator and early researcher in this field, suggests that two steps are necessary to help students modify their preconceptions. First, students need to be encouraged to articulate their ideas and to use them to make predictions. Second, students are encouraged to make explicit comparisons among their preconceptions, accepted scientific theories, and convincing empirical observations, to see which theory makes the most consistent and accurate predictions. Arons (1977) and Fuller, Karplus, and Lawson (1977) advocate similar strategies. It is much easier to produce convincing empirical evidence for beginning students in areas such as mechanics than it is in many areas of biology. For this reason, reported strategies for producing conceptual change in biology are especially valuable.

The conceptual change strategies that have been devised thus rely heavily upon giving students hands-on or computer-based inquiry learning experiences which will produce anomalous results with respect to naive theories but give expected outcomes with respect to the scientific theory. It is important to recognize that the sudden reconfiguration of an individual's mental model, such as that which occurs in a flash of insight, is the exception rather than the rule. Usually, a learner's naive theory is retained while a new and contrasting theory is being assimilated. The two competing theories then coexist in a student's mind and their relative strengths vary according to such variables as context. A student may, for example, give a scientific response on an exam but revert to the more familiar explanatory model in a hands-on situation. The process of change is extended and is not always completed. The final step, accepting and committing to the new model, allows the old model to gradually fade away.

Posner, Strike, Hewson, & Gerzog (1982) have described the steps in conceptual change in the following way. In order for conceptual change to occur, learners must first become dissatisfied with their existing conceptions and then must find the new scientific conception being offered to them to be intelligible, plausible, and fruitful. Students go through three phases in acquiring new ideas (Strike & Posner, 1985). At first they barely understand the ideas. Then they develop a somewhat more complete understanding. And finally, they accept and become committed to the new idea.

Chinn & Brewer (1993) found seven types of responses to anomalous data among students, only one of which was the desired "change of the naive theory." The other six responses were: 1) ignore the anomalous data, 2) reject the data, 3) exclude the data from the current theory, 4) hold the data in abeyance, 5) reinterpret the data, and 6) make peripheral changes to their naive theories. This underscores once again, how difficult it is to change a well-established preconception.

Linder (1993) argues that the conceptual change learning model must routinely include examples from every day life. Cobern (1996) further suggests that science needs to join with other disciplines to make large scale changes in people's world views and background assumptions so they are more inclined to comprehend science. Hopefully, we are beginning to move in these directions, as suggested by E. O. Wilson's (1998) book on *Consilience: The unity of knowledge,* which proposes the coming together of the social and natural sciences. In a somewhat different but related vein, Pinker (1994) has suggested that all learning involves interactions between the environment and many innate, hard-wired learning modules in the brain, similar to the language module proposed and demonstrated by Chomsky (1975, 1980, 1988). According to Pinker, these innate modules function in the assimilation, interpretation, and learning of different kinds of information.

For more information on conceptual change, readers may wish to refer to a special issue on conceptual change in the physical sciences (Vosniadou & Saljo, 1994). Another interesting overview is provided by a comparative meta-analysis of instructional interventions in reading education and science education (Guzzetti, Snyder, Glass, & Gamas, 1993). This meta-analysis reports that *effective procedures had a common element of producing conceptual conflict.* Smith, Blakeslee and Anderson (1993), studying teaching strategies associated with conceptual change in middle school, concluded that use of conceptual change strategies by teachers was associated with higher student performance on tests designed to assess conceptual understanding, but few teachers could implement these strategies with training alone. They needed appropriately designed curriculum materials.

For readers who wish to pursue the conceptual change literature in specific areas of interest, a few recent papers may provide helpful starting points.
- *for teacher learning* – Tillema & Knol, 1997; De Jong & Brinkman, 1997; Thorley & Stofflett, 1996; Morine-Dershimer, 1993;
- *for student learning* – Dalton, Morocco, Mead, & Tivnan, 1997; Windschitl, 1998; Stofflett & Stoddart, 1994; Wade, 1994; Basili & Sanford, 1991;
- *using analogies* – Mason, 1994; Dagher, 1994; Morine-Dershimer, 1993;
- *using computers* – Windschitl & Andre, 1998; Horwitz & Barowy, 1994;
- *in biology/chemistry* – Songer & Mintzes, 1994; Basili & Sanford, 1991.

The conceptual change literature attests to the convergence of thinking in the last quarter century among a wide array of disciplines related to education and learning.

USING MAPPING TOOLS TO TRACE CONCEPTUAL CHANGE

In considering the nature of meaningful and mindful learning in a previous chapter, we saw that such learning typically consists of adding new knowledge through a network of links to the learners' existing knowledge. With respect to misconceptions, however, we have seen in this chapter that some portion of the students' existing knowledge structures must be dismantled in order for the scientific view to find its proper place. The mapping tools described in this volume are highly adaptable to instruction designed to facilitate and document this process (Figure 4.1).

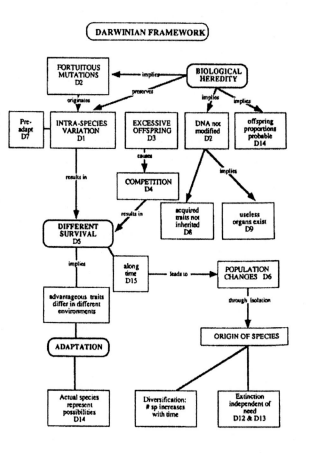

Figure 4.1. Neo-Darwinian view of natural selection. Adapted from Aleixandre (1996). The relationship indicated by the unlabeled lines might mean "includes."

Concept maps, SemNet® semantic networks, and circle diagrams (all introduced in Chapter 1) are preeminently tools designed to help the user clarify and make explicit the structure of relationships present in cognition. Each of these tools approaches the cognitive landscape with somewhat different capabilities, much as tending a garden requires tools of a variety of sizes and shapes. Mapping tools have in common, however, the expression of perceived relationships among concepts, and so are uniquely qualified to elucidate the differences between alternative conceptions and the scientific conception. Mapping is a technique that is often used to compare students' mental models with the corresponding scientific models (Fensham, Garrard, & West, 1981; West, Fensham, & Garrard, 1985; Gordon, 1989, 1996). For example, the scientific or Neo-Darwinian view of evolution that was summarized in Table 4.1 is presented again for comparison (Figure 4.1). Do you find one presentation easier to assimilate than the other?

As another example, we examine Jon's understanding of photosynthesis described in the opening vignette (Figure 4.2).

Jon's mental model of photosynthesis may be compared with the scientific view as summarized by Professor Joseph Novak (Figure 4.3). Maps such as these help researchers to analyze individual mental models for diagnostic or other purposes (West & Pines, 1985). They help to pinpoint and clarify the areas of difficulty.

In the foregoing examples it is possible to see the potential utility of mapping tools as devices to help students make the leap to the scientific view of natural processes and events (Fensham, Garrard, & West, 1981; Fisher, 1991; Gorodetsky & Fisher, 1996). A map provides an arena in which ideas can be examined and manipulated. In the authors' view, conceptual change is most effectively facilitated when students are engaged in representing their own biology knowledge as they are acquiring it. These activities hold one's ideas in place and so promote the three Rs associated with meta-learning or metacognition: *review*, *reflection*, and *revision*. Gravett & Swart (1997) assert that knowledge mapping guides, promotes, and assists in knowledge construction, helps monitor conceptual change, and identifies persisting misconceptions.

Mapping thus serves two distinctly different purposes. One is to capture what a particular individual is thinking at a given point of time and in a given context. This is often referred to as the individual's structural knowledge (Jonassen, Beissner & Yacci, 1993). The mapping captures one snapshot of the more dynamic thinking process occurring in working memory, and also reflects the organization of ideas in an individual's long-term memory. Mapping can also be used as a tool for constructing new knowledge. In this case it serves as a memory extender, holding one's ideas and freeing short-term memory for use in reflecting upon and revising those ideas. McAleese (1998) suggests that the tension between these two purposes is important in understanding the impact of mapping.

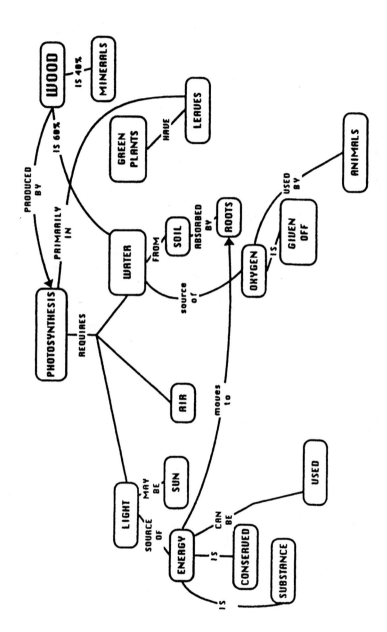

Figure 4.2. Jon's naive view of photosynthesis. Jon is a middle school student. Excerpted from the Harvard/Smithsonian program, Why Are Some Ideas So Difficult? (Schnepps, 1994). Created by Dr. Joseph Novak.

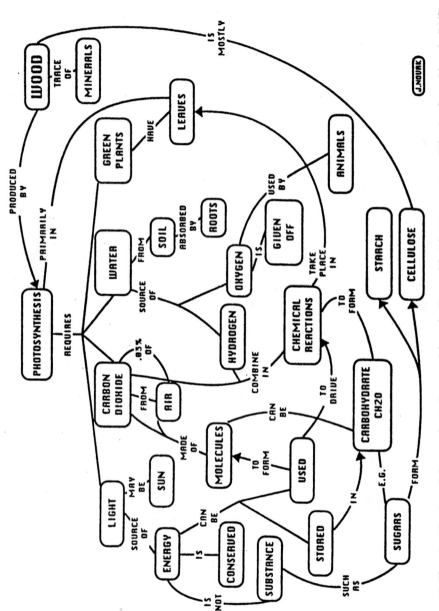

Figure 4.3. The scientific view of photosynthesis. Excerpted from the Harvard/Smithsonian program, Why Are Some Ideas So Difficult? (Schnepps, 1994). Created by Dr. Joseph Novak.

USING MULTIPLE CHOICE TESTS TO MONITOR CONCEPTUAL CHANGE

Misconceptions in physics have been more extensively studied than those in other fields. Naive conceptions regarding force and motion have been studied by investigators for at least two decades. Building on this research base, a *Concept Force Inventory* test was developed by Hestenes, Wells, & Swackhamer (1992) and is further described by Hestenes & Halloun (1995). The test is multiple-choice. What makes it unique and powerful is that it incorporates naive conceptions as distracters. These naive ideas look so implausible to physics instructors that they tend to think that the questions are ridiculously simple. They expect their students to earn high grades on it. Physics instructors are shocked, then, when they discover that the average score for their class is 40 to 50 per cent. Because of this surprise factor, the *Concept Force Inventory* has gotten the attention of many physics professors and high school physics teachers. Hundreds of high school and university physics instructors throughout the country are now using the *Concept Force Inventory* test as a benchmark of their teaching effectiveness (or lack of it).

Similarly, a two tiered multiple choice test for conceptual understanding of osmosis and diffusion has been developed in biology (Odom & Barrow, 1995). Using this test, Christianson & Fisher (1999) determined that nonmajors in a small inquiry-based biology course developed a significantly deeper understanding of osmosis and diffusion than nonmajors in large lecture courses. Another test for conceptual understanding in biology has been developed by Haslam and Treagust (1987). Although these biology tests are quite effective, they are not likely to have the same impact on the field of biology that the Concept Force Inventory has had in physics, in part because biology instruction is much more variable than mechanics instruction, and in part because mechanics more readily lends itself to clear representations of the scientific and naive ideas.

Creating a multiple choice test using misconceptions requires a lot of research to elucidate and understand the dimensions of those conceptions, as well as extensive testing and revision of the items. Not surprisingly, most multiple choice tests do not tap into common alternative conceptions. Instructors generally do not include naive conceptions as distracters because they are generally not aware that such conceptions exist. Commercial testing firms design test items to conform to certain desired statistics; items with very potent distracters are often discarded.

NAIVE CONCEPTIONS AND THE CURRICULUM

According to the cognitive conflict model of conceptual change (Posner et al., 1982), the teacher must arrange or present a body of evidence that cannot be accounted for by the student's naive theory. At the least this process should precipitate in the student a willingness to entertain new hypotheses. The difficulty of this stage in the instructional process, however, should not be underestimated. Students are very adept at finding reasons that explain away the evidence and allow their existing understanding to remain intact (Chinn & Brewer, 1993), as previously mentioned.

Thus this stage of instruction necessarily involves a major commitment of time. Class discussions and sharing of data among individuals or groups in the class also help, since it is more difficult for a student to deny results if all students made the same observations. Challenging a misconception may also require a graded series of experiments (bridging) to be sufficiently convincing. The goal is to teach for deep understanding of key ideas rather than for superficial coverage of many undifferentiated ideas.

Moody proposes the possibility of an entire strand of the science curriculum organized around key moments of insight. Each year of study beyond the first or second grade could include a focus on two or three topics known to be associated with significant misconceptions. After systematic exposure to a graded series of such topics, students may begin to acquire experience in the kind of cognitive transition required in such cases. Somewhat similarly, after a time teachers may acquire expertise in the effective presentation of such topics. An "insight curriculum," in short, may yield unexpected dividends not only in the increased ease of students' acquisition of difficult concepts, but also in acquiring metacognitive strategies for learning generally – that is, meta-learning.

Meta-learning focuses upon students' assessment of the quality of their own understanding. According to White and Gunstone (1989), *systematic incorporation of meta-learning into the science curriculum require a sustained commitment on the part of all concerned, including an awareness by the teacher that understanding per se must be recognized and rewarded.* This principle stands in stark contrast to the widespread emphasis upon curriculum "coverage" that has been characterized as "the bane of meaningful learning" (Wandersee et al, 1994). Learning to assess the quality of one's own understanding may be facilitated and supported by the appropriate use of knowledge mapping tools.

We began this chapter by looking at the common misconception, "How can carbon dioxide account for the bulk of weight in plants since it (CO_2) has no weight?" This prevalent misconception about air interferes with comprehension of many scientific processes, not just photosynthesis. Critical conceptions (protoconcepts) such as this one, which interfere with learning of numerous higher level topics, could be effectively addressed in elementary school with appropriate curriculum materials. This would help to lay a solid foundation for learning higher level concepts in later years such as how airplanes fly, how balloons rise, and what happens during state changes.

CONCLUSION

Naive conceptions are the product of each individual's efforts to make sense of the world. They are necessarily created on the basis of limited data combined with a lack of familiarity with the relevant scientific concepts. Over time, a complex pattern of ideas and events become linked to an individual's naive theory in such ways as to influence both the thinking and the behavior of that individual. For these reasons, alternative conceptions are described as being "deeply embedded in an individual's conceptual ecology" and they are highly resistant to being taught away. In fact, emotional stress is often associated with the discovery that one's naive theory is wrong.

Compelling evidence is usually necessary but not always sufficient to transform one's thinking from a naive idea that seems intuitively obvious to a scientific idea that is much less obvious. To shift from a naive to the scientific view, an entire constellation of known, existing facts must come to be seen in a new configuration. Conceptual change strategies are frequently used to achieve this end. They involve engaging students in inquiry learning experiences in which their naive theories lead them to make incorrect predictions. Learners go to great lengths to maintain their naive theories and reject their observations when the data is anomalous. A flash of insight sometimes produces the resolution of a conflict between the naive and scientific views. Such moments can bring genuine pleasure to both student and teacher. Multiple-choice items that include common misconceptions as distracters can be powerful tools for measuring understanding.

In sum, the study of misconceptions shows that the essence of what is meaningful is not the acquisition of individual, isolated names or bits of data, but rather the perception of relationships among a network of concepts and ideas. Meaningfulness, in other words, appears to reside in a coherent pattern of structures or events. Pagels (1988) describes it as a matrix of meaning (also see Chapter 9). It is this kind of pattern, rather than any particular piece of information, that the mapping tools described in this book are uniquely designed to capture and to represent .

KATHLEEN M. FISHER

CHAPTER 5

Meaningful and Mindful Learning

Real Life Can Promote Meaningful Learning!

Susan's mother, her two sisters, her aunt, and her aunt's daughters had contracted breast or ovarian cancer and three of them, all less than 45 years old, had succumbed to their diseases. For these reasons, Susan decided to have her breasts removed prophylactically. However, cancer researchers had just identified a molecular marker associated with the gene for breast cancer in Susan's family, known to them as "Family 15." The researchers hadn't thought about sharing their findings with the family until they heard about Susan's plans for surgery.

Members of Susan's family had come to believe that a breast cancer gene was being passed from mothers to daughters. Susan thus assumed she would follow in her sisters' footsteps. However, the researchers informed Susan that she didn't require surgery because she did not have the breast cancer gene. Without realizing the bomb they were dropping, they explained that 50% of all family members, males and females alike, would have this autosomally linked gene.

The many family members who had thought they were exempt from the cancer plague went into shock. Anna and Adrienne, two daughters of Susan's Uncle Doug, had assumed their father did not have the gene and thus neither did they. However, they learned within a period of less than 3 intense weeks that a) they may have the breast cancer gene, b) in fact, they did have the breast cancer gene, and c) they not only had the gene, but mammographs revealed that they also had breast cancer! Their previously secure worlds turned topsy-turvy. At the same time, they realized that their newfound scientific knowledge probably saved their lives. (Waldholz, 1997)

This vignette illustrates a mother to daughter theory of inheritance invented by a family under duress. The theory adequately accounted for the cancer cases they observed in their own family during a relatively short period of time, but the data were limited and insufficient. Under dramatic circumstances, family members were informed that the scientific theory was quite different from their own. Compelling evidence (in the form of the unexpected presence of breast cancer in two young women who thought they were safe from the scourge) supported the scientific theory. All 39 family members not only had to discard their "naive conceptions" (described in Chapter 4) and assimilate the new scientific ideas, but they also had to generate new inferences about appropriate ways of managing their lives.

Real life has a way of imposing meaningful learning on us in a highly persuasive manner. Learning and retention are generally increased when adrenaline levels are higher, as in these life and death situations. The classroom is a bit different, however. This chapter looks at the problems of achieving meaningful learning in biology classrooms.

WHAT IS LEARNING?

Learning can be a lot harder than simply absorbing new knowledge. Learners' prior knowledge and background assumptions can present major obstacles. Carefully selected hands-on experiences can serve to challenge such background assumptions and bring new understandings. Such science activities are not an end in themselves, but rather a means to an end – to develop understanding of scientific ideas. In this chapter I aim to clarify and make explicit what we mean by "understanding of scientific ideas," "meaningful learning," and "mindful learning."

Much has been discovered about how people learn in the past few decades, due in part to a convergence of theory and empirical research from many different fields. These findings seem strong because different researchers in different fields using different methodologies have come to similar conclusions. The reform movements currently sweeping educational communities at all levels, especially precollege (briefly described in Chapter 1), are attempting to bring some of this knowledge into the classroom. The goal is to generate the mirror image of how to learn – namely, how to teach.

MINDFUL LEARNING

The *processes* of mindful learning lead to meaningful understanding (Langer, 1989, 1997; Murray, 1997; Gagne, 1977). Mindful learning refers to the ways in which we function during the learning process.

The basic idea is that fluid, flexible thinking boosts our learning ability. Langer encourages us to experiment and to play with information, looking at it from different perspectives, making use of multiple examples, and exploring how the meanings of a given set of information change in different contexts. She identifies seven myths or false attitudes (Langer, 1997, p. 2) that are embedded in the educational system and that stunt students' growth and interest in learning. They are reviewed below.

First, many in education believe that the basics should be so well learned that they become second nature. This is incorrect, says Langer. Drilling in the basics leads to overlearning or learning without thinking – the automaticity described above. Does it make sense, she asks, to freeze our understanding of a skill before we try it out in different contexts and adjust it to our own strengths and experiences? One of the studies performed by Langer and her colleagues found that pianists who learned by varying their playing style performed more competently and creatively than those who learned to play strictly through repetition.

Second, educators think that paying attention means staying focused on one thing. This myth, according to Langer, fails to recognize the value of novelty in holding our attention. Her studies show that varying the target of our attention, whether it is a visual object or an idea, improves our memory of it. In one study performed with

Martha Bayliss, groups were instructed to read short stories. The "mindful" groups were instructed to vary aspects of the story such as to read from different perspectives, consider different endings, etc. The "focus" groups were told to focus their attention on certain fixed aspects of the stories. The control groups read without any specific instructions. When participants were asked to list all they could remember from the story they just read, the mindful groups remembered significantly more details than the others, even though they had the most to think about.

Third, conventional education buys into the idea of "work (learn) now and play later." Langer claims, however, that learning itself can and should be fun. She feels the fun is lost when ideas are removed from their contexts and when learning is evaluated and graded. This shifts the reward from the innate pleasure of learning to the pleasure of getting a desired grade (or the fear or disappointment of not getting the desired grade). The innate pleasure of learning, she says, comes from making finer and finer distinctions between things.

Fourth, rote memorization is prevalent in education, but Langer sees memorizing as a way of taking in information that is personally irrelevant. Rote learning is usually undertaken for the purpose of performing on an evaluation, not to achieve understanding. It is analogous to the twist that occurs in the courts as lawyers set out to win a case, not necessarily to find justice. Langer feels that one way to reduce rote learning is to encourage students to make information personally meaningful.

Fifth, memory is essential to living in the world. It provides the basis for our expectations, actions and safety precautions (e.g., don't put your hand on a hot stove). But, says Langer, forgetting can have its benefits, especially in the opportunities it provides for rethinking ideas in a new context.

Sixth, teachers often act as if intelligence consists of knowing facts. This is not the case, says Langer. Intelligence consists of thinking flexibly and looking at the world from multiple perspectives. This theme, so relevant to biology, has been elaborated by Spiro and colleagues in their cognitive flexibility theory, a theory of knowledge acquisition in ill-structured domains (e. g., Spiro, Coulson, Feltovich, & Anderson, 1988). Although Spiro developed cognitive flexibility theory to describe biology learning by students in medical school, I find it provides an excellent model for teaching nonmajor biology as well (Fisher & Gomes, 1996a).

I believe that when teaching nonmajors or majors who will be working in other fields, emphases on the "big picture" are important. Details can be obtained on an as-needed basis in the future. At the same time, detailed facts are important for those who will be working in the domain. As mentioned in Chapter 3, content knowledge about a domain is a *major determinant* of problem-solving performance in that domain. In studies of two disparate domains (mathematical vectors and using a video tape recorder), Gordon and Gill (1989) found that subjects' interconnected content knowledge, mapped in conceptual graphs, predicted *85 to 93%* of an individual's ability to solve problems in those domains. Missing concepts or missing links caused problems with performance. These studies and related research indicate that teaching isolated facts is largely useless, while prompting learners to construct a coherent and interconnected set of ideas about a domain is productive and worthwhile.

Langer's seventh point is that many teachers believe there are right and wrong answers. Langer disagrees with this belief, as do most constructivists. Science aims to produce the best model of the world at any given time; it is not necessarily the "right model," the "only possible model," or the "truth." There is awareness among scientists that any theory or observation may change or be replaced in the future, either by generation of new empirical data or by conceptualization of an even more satisfactory and powerful theory. Thinking that we have the "right" model leads to rigidity and fixedness, whereas thinking that what we have is currently the "best" model can lead to flexibility, openness, and continued willingness to question.

A key message that runs throughout Langer's discussions is that students must become motivated to learn (learning can be fun) and that students must take responsibility for their learning. Given its important role in learning, it seems that increasing student motivation to learn should be our number one priority.

MOTIVATION

In studying learning, Rumelhart and Norman (1978) observed that motivation outweighed any cognitive variables they were able to measure. Likewise, Dubin and Taveggia (1969) found that student motivation was a more powerful determinant of learning than any change in teaching strategy. Some steps which are known to increase motivation are:

- giving each student a voice in the class,
- respecting each student's input;
- allowing students to pursue their own questions;
- encouraging students to work in groups and discuss their ideas among themselves;
- creating opportunities for students to create and test their own explanatory models;
- giving students an opportunity to demonstrate their knowledge to others through publication or presentation; and
- providing tools which can sustain student analysis and discussion. *Enhancing student motivation often entails reducing emphasis on learning the facts and increasing emphasis on learning scientific processes.*

MEANINGFUL LEARNING

Ausubel (1968, pp. 37–38), a psychologist who spent his lifetime thinking about learning, describes meaningful learning in this way. The essence of the meaningful learning process is that ideas are related in a substantive (nonverbatim) fashion to what the learner already knows. Each new idea is connected to some existing relevant aspect of an individual's mental structure of knowledge (for example, an image, a meaningful symbol, a concept, or a proposition). Meaningful learning requires two conditions. First, the learner must be motivated to learn in a meaningful way, and second, the material being learned must be inherently meaningful and accessible to the learner. A third condition is that there be sufficient time for meaningful learning to occur, since learning is an effort- and time-demanding process.

Basically, learning involves a number of steps including *perceiving* the world and information in the world, *interpreting* that information, *encoding* it somehow in the mind, *retrieving* it as needed, and then *applying* the information in various contexts. Each of these steps is briefly discussed below.

1) Perception

Our perceptions are limited by our particular perceptual hardware. We cannot "see" like a satellite camera, measuring color or density differentials, nor like an eagle, spotting a small animal on the ground from high in the air, nor like a bee, taking in the ultraviolet spectrum. The world we are able to know directly is constrained and molded by our perceptual hardware.

Since our perceptual limitations filter and define our world, we can never "know" the world absolutely and totally. "Right answers" are elusive. Yet science as a "search for truth" has been a popular conception among science teachers for years. As Langer says, science is often taught as if there is a "right" answer to each question, and the students' job is to memorize those facts or truths about the world.

But this is not how science is actually conducted. Scientists strive to construct the best possible *model* of the world at any given time. They constantly evaluate their models and assess which one is best in terms of its ability to explain, to predict, and to account for many different observations. "Facts" are not necessarily truths but rather well established records of objects or events that are widely accepted to be correct, at least for the time being. In science, a prevailing model can be replaced with another at any time, if the newer model is more powerful and satisfactory. The replacement process can be painful for individual scientists in the "out" group, those who are still attached to the old ideas, especially where large conceptual revolutions are involved (Kuhn, 1970).

Teaching science as if it consists of facts alone is self-defeating, in part because the facts keep changing. Students need to understand that the scientific way of knowing is based upon systematic study of objects and events combined with the construction of models to explain and predict (although prediction is not often possible in the retrospective sciences). Models are tested under a variety of circumstances and by many different scientists. Creation of scientific knowledge is thus a collaborative venture. The public is often confused when they hear conflicting beliefs and claims by different scientists, but such disparate viewpoints are a natural part of a group knowledge-building effort that relies on individual ingenuity, collaboration, and competition. When a particular knowledge claim is challenged, its supporters are prompted to find even more convincing evidence to support their point of view, and so science advances.

A surprisingly effective way for students to learn about the scientific process *and* to develop a fairly deep understanding of science content is to read a good popular book on a subject. In my experience, biology nonmajors who read and discuss *The Beak of the Finch* (Weiner, 1995) while also completing a series of related lessons in biology (Fisher, 1999), come to deeply understand many aspects of evolutionary theory and they gain insight into the nature of microevolutionary research – even if they have religious reservations at the beginning. Likewise, Watson's (1991) *The Double Helix* really challenges students' naïve beliefs about how science works. Furthermore, students enjoy reading these well-written books with good story lines. I now use them instead of textbooks, and consider this one of many wake-up calls for textbook publishers.

2) Interpretation

Our perceptions are not only constrained by our neural hardware, but are also informed by our cognitive frameworks. We build up expectations about the world and then tend to see what we expect to see, even when it isn't really there. The Science Media Group at the Harvard/Smithsonian Center for Astrophysics has produced a series of videotapes in which they interview students about certain scientific concepts (Schnepps, 1997a, 1997b). In one interview, Jennifer expresses her belief that she can see in the dark once her eyes adjust. She continues to express this belief even after sitting for six minutes in a completely dark room in which she could see nothing. Rather than give up her theory, she chooses to assume that more time is needed for her eyes to adjust – maybe as much as a year (Science Media Group, 1997). As Dr. Stoddard, an educational psychologist at the University of California, Santa Cruz, points out in the video, we like to think that "seeing is believing", but in fact, the reverse is often true, *"believing is seeing!"*

In a similar vein, if an experiment doesn't come out as students expected, their usual conclusion is that "I must have done the experiment wrong," rather than, "My theory must be incorrect." They are attached to their beliefs and resist giving them up. In fact, Chinn and Brewer (1993) showed that students have six different strategies for rationalizing away anomalous results, compared to just one strategy for assimilating the results into their mental models.

Not only are students attached to their naïve theories, but they are likely to misinterpret new information many different ways. Problems can arise, for example, when a term has multiple meanings. As one illustration, the terms "population", "community", and "habitat" each have specific scientific meanings as well as loosely defined popular meanings. Furthermore, biologists are likely to slip back and forth between the two meanings, as in the "preying mantis population" (specific) and "the insect population" (general). There is also a tendency to use the name of an organism such as the "owl" to refer to two very different things: an individual owl and a population of owls. Experts manage to communicate with each other in spite of the lack of discrimination in terminology, but for many students it is a struggle.

Biology also has a lot of "historical" baggage, instances where the language hasn't caught up with modern understandings. One example is the undifferentiated use of the word "chromosomes" to designate multiple structures, ranging from the extended, unreplicated chromosomes of early interphase, to the condensed, replicated chromosomes of metaphase. To the condensed unreplicated daughter chromosomes of anaphase, with all the intermediate forms in between. The lack of semantic differentiation poses severe and unnecessary challenges to learners, especially when the undifferentiated term is used in describing such events as mitosis and meiosis, as is the case in many textbooks. This peculiar situation presumably arose because chromosomes were first studied by individuals with relatively little knowledge about their structure and functioning. Why the terminology hasn't advanced along with the deepening understanding of the subject is a tribute to the power of history and inertia.

Life has fuzzy edges, which makes precise definitions of many biological ideas rather elusive. For example, what is a "<u>predator</u>"? We once found nine different definitions for this term in introductory biology textbooks. But having two or more meanings for the same word is only the most obvious level of (mis) interpretation. Unfortunately, one of the most serious consequences of transmission instruction is that the instructor rarely ever becomes aware when problems in communication exist, even though learners' interpretations of new information are always colored by their background knowledge.

Research shows that once a mental model becomes established in an individual's conceptual psychology, it can be highly stable. New information is molded by the learner to fit into the existing model. During the past twenty-five years or so, more than 3200 studies have been conducted to elucidate students' "naive ideas," "alternative conceptions," or "misconceptions" about science (Wandersee, Mintzes, & Novak, 1994; also see Chapter 4). Students around the world share many common misconceptions. These conceptions are typically held by a significant proportion of students in any given class, and most critically, they cannot be "taught away" simply by *telling* (that is, by lecture or transmission instruction). And of course, if an instructor has no idea that they even exist, she or he can't begin to address them.

Thus, interpretation and misinterpretation are key events in learning. This may account in part for early reports that students learn more in internet courses than in traditional classrooms, because we tend to be more careful and precise with written than with spoken words. Collaborative workgroups with plenty of structured opportunities for discussion also significantly reduce the frequency of misinterpretation.

3) Encoding

Like perception and interpretation, encoding is susceptible to many errors. We encode what seems at the time to be key features of a situation — not the entire experience or observation. Our minds do not behave like cameras. When we recall a particular experience or observation, our minds reconstruct our memories rather than retrieving precisely what was initially perceived. For each of the details that weren't recorded initially, memory inserts "default" settings (for example, blue sky, green grass), many of which may be incorrect for the actual situation we observed. This is why multiple witnesses of an accident rarely agree on all the circumstances (Bourne, Dominowski, & Loftus, 1979). They may have witnessed the same event, but they saw it from different perspectives, paid attention to different elements, and reconstructed their memories somewhat differently. Even if they encoded a feature correctly at the outset, it can be changed over time through subtle suggestions or confusions (Bourne, Dominowski, & Loftus, 1979).

4) Retrieval

Retrieval can also be challenging and it gets more difficult as we grow older. It may take hours or days to retrieve the name of a person we have just seen, or of an author whose book we want to find again. And retrieval may be partial as in the "tip of the tongue" phenomenon (Brown & McNeill, 1966), where one can remember what letter(s) a word begins with and perhaps how many syllables it has, but not the word itself.

Experts can retrieve knowledge in their field and can apply it more effectively than can novices. This is due in part to the fact that experts typically have many more connections to each idea than do novices and think about them quite differently (e.g., Chi, Feltovich & Glaser, 1981). Ideas are thought to become more prominent in memory due to a phenomenon called *spreading activation* (Collins & Loftus, 1975; Norman, Rumelhart, & the LNR Research Group, 1975; see also Chapter 9). According to this theory, when an idea such as "electron transport chain" is recalled, its activation level is increased, and the activation spreads to related ideas such as "mitochondrion," "matrix," and "cristae." In general, the more connections or pathways there are to an idea in memory, the greater the chance of giving that idea a boost either by direct activation or indirectly through spreading activation from related concepts. Experts also tend to develop more systematically organized, well-constructed hierarchies. These structures seem to aid retrieval because the memory easily "travels down" these well-worn pathways (Reif & Heller, 1981; Reif, 1983).

5) Applying Knowledge in New Situations

Applying knowledge to new situations is always challenging. Even experts can falter in this area. In general, near transfer has greater likelihood of success than far transfer. That is, the more similar the new situation is to the learned situation, the easier it is for an individual to apply their relevant knowledge. The theory of situated cognition suggests that the best way for students to learn a topic is to have the knowledge embedded within the context in which the knowledge is most likely to be applied (Ragoff & Lave, 1984). Lave, Murtaugh, & Rocha (1984) found, for example, that individuals were able to do math with many fewer errors in the familiar grocery store setting than in artificial or academic settings. This line of research suggests that people will perform better in familiar contexts. There appears to be no good theory for promoting people's ability to transfer what they know to new, unanticipated situations other than practice in performing such transfers.

Automaticity is a powerful feature of performance in well-rehearsed domains (Anderson, 1983). Automaticity allows a cognitive task to be performed subconsciously, without taxing conscious thought. It is what allows us to drive, listen to the radio, and talk on the phone at the same time. However, automaticity also has a down side. Once a cognitive procedure becomes automatic, it can be difficult for an individual to know exactly how she does what she does and to explain it clearly to someone else. Many experts have this problem in their areas of expertise. It can also be challenging for the individual to change the automated procedure because it normally occurs beyond conscious control.

In summary, learning for understanding implies that the learner has made sense of the situation or material. Sense-making involves adequate *perception*, *interpretation*, and *organization* of ideas in suitable ways, and *encoding* in long term memory with appropriate connections to prior knowledge. The learner needs to be able to *retrieve* the ideas from memory as needed, and *apply* the ideas in multiple contexts. Cognitive supports for learners, including allowing sufficient learning time and providing opportunities for interactive conversations about the topic, can make a big difference in students' success in learning biology. Mapping and semantic networking can help students develop the cognitive and metacognitive skills they need to master a complex subject and to resolve contradictions among competing ideas (Figure 5.1).

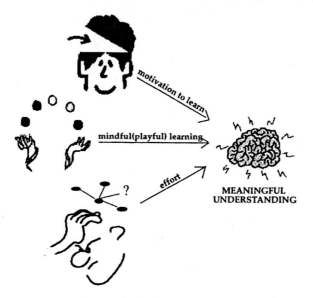

Figure 5.1. Factors that facilitate learning. By L Becvar.

ATTITUDES ABOUT ERRORS

Asian students regard errors as a natural part of learning (Stevenson & Stigler, 1992). In fact, errors are appreciated as a signal that a learner must be persistent and work harder at acquiring understanding. In contrast, when American students make errors, they are inclined to assume that they do not have an aptitude for the subject and should avoid it (Stevenson & Stigler, 1992). The American belief that ability is innate thus tends to be self-defeating, while the Asian belief that ability can be acquired through hard work is empowering.

ROTE LEARNING

Another way to examine meaningful learning is to look at what it is not. Rote learning or memorization is at the opposite end of the spectrum from meaningful learning. Learning by rote typically produces isolated, superficial and temporary knowledge, as when one remembers a telephone number long enough to dial it or a street address long enough to find it. Such memorized information is retained primarily through rehearsal and doesn't last much longer than the rehearsal effort. In fact, if a person is interrupted before finishing the task, chances are she/he will have to look up the number or address again.

Sometimes, however, memorization serves as a starting point and provides scaffolding for the meaningful learning that follows. This is seen in the learning of such things as multiplication tables, spelling conventions, and rules of grammar. When rote learning is both the beginning and the end of learning about a topic, however, it is safe to say that the learner doesn't know much about that topic. Knowledge acquired by rote learning is largely inert and inaccessible for problem solving, pattern recognition, and other mental tasks.

Ausubel (1968, p. 38) examines some of the forces that lead students to resort to memorization. Memorization occurs when students discover that correct answers, which lack verbatim correspondence to what they have been taught, receive no credit from certain teachers. Students also resort to rote learning when they have a high level of anxiety or they experience chronic failure in a given subject, resulting in a lack of confidence in their ability to learn meaningfully. And students engage in rote learning when they are under pressure to conceal, rather than admit and gradually remedy, lack of genuine understanding.

Unfortunately, memorization is probably the most widespread learning practice in schools, thanks to the combined effects of transmission (lecture) teaching, absence of dialogue with and among students, ditto sheets for practice exercises, and multiple choice testing. In addition, Stevenson & Stigler (1992), in studying American, Japanese and Chinese classrooms, found that American students worked alone much more frequently. Asian students, in contrast, are given many more opportunities to talk about a topic in whole class and small group discussions.

American students discover that they can utilize memorization with great success in the short term, performing well and getting good grades. But they also eventually discover that they are disabled in the long term, because memorization eventually collapses under its own weight. Table 5.1 compares some aspects of rote and meaningful learning.

Table 5.1. Rote versus Meaningful Learning

Rote Learning	Meaningful Learning
isolated	interconnected
arbitrary	coherent
irrelevant	relevant
nonfunctional	functional
quickly forgotten	long-lasting

PERSONAL KNOWLEDGE CONSTRUCTION

Constructivism is a theory of learning first developed in this century by Kelly (1955) and made popular in recent years in part through the efforts of von Glasersfeld (1984, 1987, and 1993). The basic idea is that each student is responsible for her or his own learning and gradually acquires new knowledge and skills by effortfully building on prior knowledge. Many science educators today agree with certain basic constructivist premises such as:

1. Cognitive knowledge must be *constructed* by the learner, as opposed to being passively received,

2. an individual's *prior knowledge* strongly influences what new ideas that individual will be able to comprehend and how she or he will interpret that new idea,

3. new ideas are understood in part through their *connections* to prior knowledge,

4. knowledge construction is *effortful,*

5. knowledge construction is enhanced by *practice, reflection, explicit analysis, and revision,*

6. knowledge construction involves a variety of *cognitive and metacognitive skills,*

7. new ideas are most productively introduced to students through observation of and interaction with *relevant phenomena,*

8. group work is valuable because *social interaction* strongly influences a powerful mechanism in knowledge construction,

9. *conversation* is important as it prompts students to convert their implicit knowledge into explicit language and to compare their ideas with those of others,

10. our model of the world can never be more than an *approximation of reality* (that is, there is no "truth" in science), and

11. knowledge is acquired gradually in part through the process of making *finer and finer distinctions* between things.

PRACTICAL SUGGESTIONS FOR PROMOTING
MEANINGFUL BIOLOGY LEARNING

Theories about learning are, to a large extent, synergistic and complementary, not competitive. They produce a strong consensus about what is important in the classroom (Table 5.2). If we could magically transport everything that we know about learning into every classroom in the world, we would take a giant step forward.

Dr. William Schmidt, Director of the recently completed Third International Mathematics and Science Study, has described the American K–12 science and mathematics curricula as a mile wide and an inch deep (see TIMSS Site Index, 1999; Schmidt, McKnight, & Raizen, with Jakwerth, Valverde, Wolfe, Britton, Bianchi, & Houang, 1996). Relative to countries whose students outperform American students, according to Schmidt, the American curriculum includes significantly more topics every semester, spends less time on each topic, and repeats topics year after year. In contrast, our most successful counterparts such as Japan and West Germany tend to teach a relatively small number of topics each semester, to spend a significant amount of time on each, and to teach each topic just once in the entire curriculum.

College biology curricula in the US are especially committed to coverage at the expense of concept and skill development. Students can rarely sink their teeth into an idea and explore it in depth. And how often are biology students given an opportunity to generate and test a hypothesis? How often are biology students asked to predict what will happen in their lab, write down their predictions, and explain why they think in the way that they do? How many biology labs engage students in reflecting on their observations, making sense of them, and explaining their mental models for why things happened the way they did?

The sheer volume of information to be conveyed keeps growing at a phenomenal rate, creating strong pressures against teaching biology for greater depth of understanding. The example of genetic diseases illustrates the kind of solution that needs to be found. So long as only half a dozen or so genetic diseases were well understood, each one could be studied in some detail. However, now that more than 5000 genetic diseases are known, in-depth study of each one is no longer reasonable, possible, or desirable. Examination of key ideas such as disease mechanisms, defenses, treatments, and patterns of heredity become important, with specific diseases serving as exemplars.

Breadth (coverage) is probably the single biggest problem confronting biology education. Biologists have a long-standing commitment to breadth in their courses, and the explosion of new knowledge pushes teachers and students to "cover" more every year. Yet research shows again and again that when complex ideas are skimmed over and not developed in a deep way, they are quickly lost from the recipients' memory banks. In contrast, research from many sources shows that teaching fewer concepts at greater depth can enhance student understanding *(less is more)*. Further, once students develop a reasonably detailed and accurate mental model of a complex process or structure, it becomes a fairly permanent part of their conceptual landscape. The idea does not need to be retaught in the next course, but rather can serve as a starting point for adding new ideas.

Table 5.2. Changing Science Teaching Strategies. National Science Foundation, undated (~1996/1997), p. 14.

Less Emphasis On	*More Emphasis On*
Treating all students alike and responding to students as a whole	Understanding and responding to individual student's interests, strengths, experiences, and needs
Rigidly following curriculum	Selecting and adapting curriculum
Focusing on student acquisition of information	Focusing on student understanding and use of scientific knowledge, ideas, and inquiry processes
Presenting scientific knowledge through lecture, text, and demonstration	Guiding students in active and extended scientific inquiry
Asking for recitation of acquired knowledge	Providing opportunities for scientific discussion and debate among students
Testing students for factual information at the end of the unit or chapter	Continuously assessing student understanding
Maintaining responsibility and authority	Sharing responsibility for learning with students
Supporting competition	Supporting a classroom community with cooperation, shared responsibility, and respect
Working alone	Working with other teachers to enhance the science program

An interesting example of what can be accomplished is provided by CGI or cognitively guided instruction (Hiebert, Carpenter, Fennema, Fuson, Wearne, Murray, Olivier, 1997). With CGI, youngsters work together in small groups to solve mathematics problems and to share their problem-solving strategies with one another after each problem is completed. The range of strategies employed by young children in solving simple addition problems is fascinating, and as students learn of strategies other than the one they used, they build up a repertoire of problem-solving skills. *With this method, first grade students have progressed to the fourth grade level by the end of their first year in school, and they remember what they learned in subsequent years* (Hiebert, Carpenter, Fennema, Fuson, Wearne, Murray, Olivier, 1997). This illustrates the *"more"* part, the gains that can be made, when we talk about *less is more.*

Recommendations for how to improve biology teaching are widely available (e.g., National Research Council Commission on Life Science, Board on Biology, 1990; National Science Board Commission on Precollege Education in Mathematics, Science and Technology, 1982; National Science Board on Precollege Education in Mathematics, Science & Technology, 1983; National Science Board Task Committee, 1986). If these ideas are incorporated into college biology curricula, we predict that college biology students will begin to acquire more fluid, flexible, and usable knowledge.

MOVING FROM EXPERIENTIAL TO CONSOLIDATED INFORMATION

Learning by doing (that is, acquiring experiential knowledge) is quite different from learning by being told. One of the obvious benefits of learning by doing is that students immediately become aware of some of the contexts and constraints involved in the experience. These important features are often omitted in teaching by telling.

In guided discovery and constructivist learning, the preferred approach is to introduce students to a new idea by exploring a phenomenon that illustrates that idea. An important element in studying a phenomenon, rarely used in "traditional" science labs, is to engage students in predicting outcomes. In order to predict, students must construct a mental model of the situation and run a mental simulation. This exercise prompts students to make their expectations explicit. It also increases their interest in the outcome, and it often reveals (to the teacher) the students' understandings of the system.

If the outcome differs from students' expectations (as is often the case with exercises selected to challenge students' assumptions), the students must attempt to reconcile the two and develop a new explanation. As noted previously, students' initial response is often to say, "I did the experiment wrong." But if most groups in the class got the same result, that result is less questionable and students must rethink their mental models of the event.

The "guided" part of "guided discovery" requires a lot of attention. Students rarely see a situation the way that a teacher sees it because of their lack of background knowledge. They need help in seeing effectively and they generally need to be prompted in various ways to focus on deep rather than surface features of a situation. Prediction is one important strategy. Probing questions introduced at various points in the lab are another. And sometimes, metaphors, analogies, or other bridging experiences are needed to help students make a transition from something learned earlier to a new idea (e.g., Lakoff, 1987; Lakoff & Johnson, 1981; Clement, 1982b, 1988).

Finally, students need both time and guidance in order to convert their "fuzzy" experiential knowledge into more explicit organized knowledge. In biology the form of knowledge organization most widely used is semantic, although it does take other forms (e.g., visual, mathematical, diagrammatic). Symbolic descriptions are essential because they allow students to talk about their observations, write about them, retrieve them easily, and compare them in systematic ways.

Yet translation from experiential knowledge, stored mostly as images and sensations, into systematic symbolic knowledge is an *effortful* process. When teachers stop a lesson at the end of a science activity, this important step of helping students to codify and consolidate their knowledge is lost, just as when instructors give students the end result (codified knowledge) without the advantage of experience with the phenomenon. It is in this knowledge *consolidation* step that mapping techniques have the most to offer.

MOVING FROM CONTENT KNOWLEDGE TO SKILLED USE OF CONTENT

Accretion (Rumelhart & Norman, 1978; Norman, 1993) or *assimilation* (Flavell, 1977) is the process of adding to one's stockpile of knowledge. As long as the learner has an appropriate conceptual framework to build upon, accretion is relatively easy, requiring little or no conscious effort. However, when there is not a good conceptual background, then accretion is slow, painful and arduous. And if the learner's prior knowledge contains a significant misconception (Wandersee, Mintzes, & Novak, 1994), addition of a new idea may be seriously distorted or blocked altogether.

Other factors can also make assimilation of new knowledge difficult. In some cases reorganization of an older knowledge structure is required, as where ideas are subsumed under a new category or rearranged within a group. In other cases, it may be necessary to revise a causal chain. This is one interpretation of Piaget's *accommodation* (Piaget's term described in, among other places, Flavell, 1977) and it is what Norman refers to as *restructuring* (Rumelhart & Norman, 1978; Norman, 1993). Restructuring is the most difficult part of learning and typically requires a significant amount of effort.

To become an expert in a given knowledge domain, one needs to acquire not only an optimized knowledge structure but also an ease and fluidity in working with that knowledge. This is *skill* development and is accomplished by what Norman describes as *tuning*. A novice initially performs a skill slowly, with conscious thought and effort involved in each step along the way. An expert can perform the same task beautifully, automatically, and without consciously thinking about it. In between are thousands of hours of practice. Norman (1993) estimates that it takes two years of full-time effort (5000 to 10,000 hours) to turn a novice into an expert. Tuning occurs with both intellectual skills such as diagnosing a patient's problem and with motor skills such as playing tennis. Further, expert behavior must be constantly retuned through practice. When an expert stops practicing, his or her skill deteriorates. Langer (1997) points out that automaticity can be harmful if acquired too early, before an optimum level of proficiency is achieved, but it has clear advantages in expert performance.

DEVELOPING HIGHER ORDER THINKING

Higher order thinking emerges as skilled use of meaningfully learned material develops. The following list of higher order thinking skills was generated by Lauren Resnick, former Chair of the American Psychological Association and former Editor of the journal, *Cognition*.

• Higher order thinking is *nonalgorithmic*. That is, the path of action is not fully specified in advance.

• Higher order thinking tends to be *complex*. The total path is not "visible" (mentally speaking) from any single vantage point.

• Higher order thinking often yields *multiple solutions*, each with costs and benefits, rather than unique solutions.

• Higher order thinking involves *nuanced judgment* and interpretation.

• Higher order thinking involves the application of *multiple criteria*, which sometimes conflict with one another.

• Higher order thinking involves *uncertainty*. Not everything that bears on the task at hand is known.

• Higher order thinking involves *self-regulation* of the thinking process. We do not recognize higher order thinking in an individual when someone else "calls the plays" at every step.

• Higher order thinking involves *imposing meaning*, finding structure in apparent disorder.

• Higher order thinking is *effortful*. There is considerable mental work involved in the kinds of elaborations and judgments required. (Resnick, 1987b, p. 3, emphasis added)

Most biologists would be pleased if their students exhibited higher order thinking skills. What they don't realize is that they can do much toward developing such outcomes. Often when teachers think about improving instruction, their minds go to changing the curriculum — changing the order in which topics are presented or taking one topic out, putting another topic in. Rarely do they think of changing the ways in which they teach or of introducing tools that support learning of complex domains. Yet research has shown again and again that we need to rethink our instructional methodology if we want to produce more competent students. And in that rethinking, it is important to keep in mind the steps that induce meaningful learning and the ways in which meaningful learning support higher order thinking (Figure 5.2).

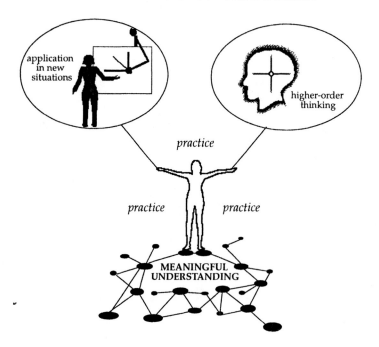

Figure 5.2. Meaningful understanding and fluidity with the material that has been mastered provides the foundations for higher order thinking and for the application of knowledge in new situations. By L Becvar.

SUMMARY

Mindful learning refers to the manner in which one approaches learning. Mindful learning involves playing with information, approaching it from multiple perspectives, and thinking about how to use and apply the information. Mindful learners maintain flexibility, openness, and an ongoing inclination to question.

Student *motivation to learn* is the single biggest determinant of whether or not a student will learn for understanding. Motivation therefore deserves considerable attention by both teachers and students, and some methods known to increase student motivation are cited above.

The process of mindful learning combined with motivation to learn typically leads to *meaningful learning* or *understanding*. These terms generally refer to the end product of learning; that is, to the state of understanding that is achieved in the learner. Learning for understanding implies that the learner has made sense of the situation or material through *perception* and *interpretation*, has organized the ideas in suitable ways and *encoded* them in long term memory with connections to prior knowledge, can *retrieve* the ideas from memory in multiple ways, and can *apply* the ideas in multiple contexts *(transfer)*.

Recognizing that we are all capable of learning is an important step. Some people may have to work harder than others to understand the same idea, which is why *persistence* and *effort* are important components in achievement.

American students today discover that they can use *rote learning* with great success in the short term, but they also sadly discover that memorization eventually collapses under its own weight.

The tenets of *personal knowledge construction* emphasize that knowledge cannot be "given" from one person to another. Each individual must construct his or her understanding of each new idea that is encountered.

Experiential learning through laboratories and other hands-on exercises can be valuable and is underused in most classes today. However, time must be taken to convert and consolidate experiential knowledge into *semantic knowledge*. Doing a science activity is largely meaningless unless time is also spent making sense of that activity. This is where mapping strategies have the most to offer.

Once students have acquired meaningful understanding, they can begin to *use that knowledge in skilled ways* and *apply the knowledge in new contexts*. The skills involved in using knowledge increase with practice, practice, and practice.

With a solid foundation of meaningful understanding, students can move into *higher order thinking*. To promote this transition, teachers must model higher order thinking skills and give students lots of opportunities to practice such skills themselves.

JAMES H. WANDERSEE

CHAPTER 6

Language, Analogy, and Biology

On the Strangeness of Learning a Language

Parents in every culture speak some form of baby talk to their infants (Blakeslee, 1997). Listen to a young U.S. mother talking to her baby: "Myee, myee, Mykelll, what a biiiiig, re-edddd, dummppp terrukkk you have there!" Janet Werker and Les Cohen are researchers at the University of Texas who study how babies learn language (Barinaga, 1997). Working with 14-month-olds, they were surprised to find that their subjects could learn to associate a particular word with a particular image – and that the babies would actually notice (as indicated by their studying the object longer) if the word was changed.

MEANING-MAKING HAS TOP PRIORITY

Babies first focus on *meaning*, rather than on the phonetic differences in the speech sounds they hear. Researchers speculate that such prioritization may be due to the fact that fine sonic distinctions really don't matter until much later, when babies' expanded vocabularies are replete with similar sounding words.

When babies start learning a language, their tiny vocabularies typically contain words that are all rather sonically distinct. Thus, it is efficient for them to disregard the fine details of word sounds when beginning to learn a language. As for the "parentese" (the accepted term for the sing-song, exaggerated way we humans enunciate the words we say to babies) found in the opening of this vignette, Patricia Kuhl and her colleagues at the University of Seattle have evidence which suggests that such parental speech actually helps babies learn the key features of vowel sounds. Some of these features are used immediately, while others pay off later when babies' growing vocabularies demand finer distinctions. The audio caricature of normal speech emphasizes key features via exaggeration – just as a political cartoon does by exaggerating a newsworthy politician's key facial features.

"English-speaking children first learn to mimic their parents' sounds, through trial and error, between the ages of three and five... [but] children need until the age of ten or so to get really good at them" (Lieberman, 1997, p. 22–27). Meaning-making comes first; speech follows. Words are, indeed, the vehicles of meaning that we use to carry our concepts – just like Michael's toy truck (in the opening vignette) can carry sand at the beach.

The situation in biology is similar. Meaning-making involves understanding a concept or idea. It often involves visualization, as in visualizing a three-dimensional cell or the dynamically moving process of protein synthesis or the notion of a niche. Learning is typically more meaningful when students are introduced to a concept first, especially if the introduction is through observation and/or experience. The associated vocabulary follows the assimilation of the idea on an as-needed basis. Unfortunately, much of biology teaching follows the opposite pattern – a word is introduced, a definition is offered, but the concept itself is never quite developed. This necessarily results in a very superficial kind of learning.

CONTEXT

As someone who has authored a number of journal articles in *The American Biology Teacher* and in *Adaptation* on the terminology problem in biology education (and coorganized a national symposium focused on it), I have long been intrigued by the role that biological terms play in learning biology. I have gradually come to think that a cost-benefit analysis ought to be conducted for the set of terms we intend to introduce in a particular biology course. The terms we choose to learn and use have implications for the kind of biology knowledge maps we construct.

THE TERMINOLOGY PARADOX

Here is the terminology paradox I see. The advantage of using a biological term instead of a common word lies in precision of communication. The gain in precision is useful to the biologist in transmitting specialized knowledge to other biologists. However, biology jargon is less useful for communicating with introductory biology students and nonmajors. Their comprehension is actually reduced when the instructor uses specialized terms. Further, the students' actual and immediate need for the terms, as well as their opportunities to use the terms, are limited. This actually results in a loss of general communication range when the students try to use the biological term in conversations with nonbiologists.

Certainly the work of life science researchers is advanced by the use of specialized terms, because what they know (and do not know) can be stated with greater specificity. Those who want to become biologists eventually need to master the terms of their field as well. However, it is not as clear that the students we want to help become biologically literate citizens need to learn as many terms up front as are usually taught – especially terms foreign to their own experiences.

The more precise and abundant the biological terms we bring into our classrooms, the fewer the students capable of understanding what we say. It is possible to describe almost everything in biology in ordinary language, without invoking biology jargon. However, this manner of speaking doesn't come easily to a highly trained specialists whose language and thought are imbued with specialized terms. It is a skill they must work to acquire. Fisher (personal communication) made this type of transition when she shifted from teaching biology majors to teaching nonmajors. She noticed that it took constant awareness, considerable effort, and sometimes deep thought to discover how to describe biological phenomena in plain English. It also took several years of practice to get reasonably good at it.

Ideally, we want our students to understand biology, yet still be able to communicate effectively with those who don't. That is, if we have to choose between the concepts and specialized terminology, it is preferable that students learn the concepts or ideas along with the ability to describe them effectively to others. They could also "learn how to learn" this secondary layer of specialized terms on an "as-needed" basis.

PREVENTING TERM-INAL ILLNESS

Can you remember having to "learn" (i.e., memorize) all of the length, mass, and volume units and combinatorial prefixes of the metric (Systeme Internationale) system before you had any compelling reason to use them or to differentiate between them? That's what it's like for students to learn biology words describing things that they haven't yet experienced, that refer to more sophisticated biological objects, events, and properties than they currently understand, and that they don't ever actually talk about in everyday life.

Those who prepare science museum exhibits are conscious of the need to prepare display labels that translate the special vocabulary of biology into something the public will want to read (even while standing) and will grasp without a substantial science background. Consider the example that follows. The first excerpt was written for the public from a research institute perspective, the second from a science and technology museum perspective (for a museum display).

1. Using a computerized digital light microscope, Rob Apkarian of the Yerkes Primate Research Center of Emory University was able to detect evidence of atherosclerosis (blockage of vessels that can lead to heart attacks and strokes) in the inside wall of a human blood vessel. The two-dimensional image, shown on the video screen and in the photograph at the right, magnifies the surface of the wall 1,200 times. The image shows early atherosclerosis – as indicated by the white blood cells (dark colored pimple-like structures) stuck to the wall of the vessel (AAAS, 1997, p. 1445).

2. THE CRIME: TRESPASSING IN THE BLOODSTREAM. A mob of tresspassers have clogged a major artery, blocking the flow of important traffic. This dangerous condition, called atherosclerosis, may lead to heart attack or stroke.

   THE DETECTIVE: THE GAME IS AFOOT. Rob Apkarian uses a computer-controlled light microscope to examine the scene of the crime: the inside wall of a human blood vessel (AAAS, 1997, p. 1445).

What term and content changes did you notice? What has been lost and what has been gained? How does this illustrate the decisions and trade-offs that must be made in communicating science to nonscientists in a museum setting?

Because many biological terms have Greek or Latin origins and are polysyllabic, they are initially difficult for students to read, pronounce, and spell. For these reasons, these terms cannot simply be assigned for self-instruction or effectively taught using a single definition, at least not if they are to have lasting meaning for the student. In addition, biology students need many more opportunities to practice using the terms, "talking biology" and "writing biology," than are provided in most lecture courses. Without opportunities to employ the language of biology that is presented to them, the language remains "dead" and dissociated in students' minds. Practice, in contrast, promotes assimilation and fluidity. It can also be useful to spend time dissecting and associating the spelling of the terms with the etymological meanings of their component prefixes, roots, and suffixes.

THE TERM-TERMINOLOGY DISTINCTION

You will note that I am talking about learning *terms* and not *terminology*. The two words are not interchangeable. Terminology (literally, the study <u>of</u> terms) is the study of nomenclature, or, it can also mean the sum total of *all* the technical terms of a science, such as biology. Since we never expect students to learn *all* the terms in a field, it is usually more appropriate to speak of learning *terms*, than of learning *terminology* – unless our focus really is on the study of terms and associated naming conventions, rather than on the terms themselves.

ANALYZING THE NATURE OF BIOLOGY TERMS

In preparing this chapter, I searched diligently for a list of biology terms that I thought offered a justifiable and representative sample of the words students typically confront in introductory biology courses. I finally settled on the list I've titled *A Defensible Basic Biology Corpus* (Appendix 6.1). This biology term list is based on the *BioA2Z A Dynamic Glossary* CD-ROM, 1996, authored by Meighan, Wan, & Starratt (1996), which is part of the CD-ROM-based *Biology Survival Kit* that I purchased from Saunders College Publishing, Philadelphia, PA 19106 USA. The list of terms I created includes the complete term list given in *BioA2Z A Dynamic Glossary*, along with 50 additional words gleaned from the 49 biology topic summaries in that CD-ROM. I thought these terms were also important but they were not included in the CD-ROM term list. The added terms are #758–#807 of Appendix 4.1. The list is centered around 51 key biology ideas and their allied concepts. The authors of the source list claim these terms are central to successfully "surviving" a biology course. Note the bleak metaphor of death that portrays biology as a threat to survival (and to fun).

THE THING-PROCESS DIVIDE

Fisher (October, 1997, personal communication) has pointed out that it may be useful to examine the *process* words used by biologists – since she has observed that a preponderance of biology terms are nouns that represent *things*. In fact, most abstract processes in biology are "reified" – that is, given noun names (e.g., vascularization). This may occur because nouns are easier to learn and to locate in psychological space than are verbs. It has been observed that children master nouns before verbs (Gentner, 1978) and so do second language learners (Rosenthal, 1996) when learning a lexicon – so there is linguistic consistency.

For the purposes of this chapter, I shall define a process as a series of steps or changes which bring about a natural or artificial (human-influenced) result, and to which we have attached a term, for example, photosynthesis. It may be helpful, at this point, to distinguish a natural process (e.g., photosynthesis) and a human-influenced process (e.g., deforestation). The latter type of process can be further subdivided into procedures, methods, and techniques – all of which are human-driven processes that vary in scope. A procedure is a general approach (or series of steps) for effecting something. A method is a basic systematic (if-then) procedure to accomplish something. A technique is a systematic (if-then) procedure for accomplishing a specific task. Note the narrowing "cone of specificity."

A PROCESS ANALYSIS OF THE BIOLOGY TERM (CONCEPT) LIST

A *process* analysis of *A Defensible Basic Biology Corpus* yielded the following results. *Process* terms (including reified process terms) accounted for about one-fourth (22.8%, or 1 out of every 4.4 terms) of all the terms in the list (185). It is not easy for novices to identify the *process* terms in biological text by their characteristics alone, because of context dependence. *Acid rain,* for example, can be either a *thing* or a *process*, depending on how it is used. Processes always involve sequences of events, rather than just objects, properties, or states.

If the list is representative of biology terms, the suffix *–ion* (or *–sion*) is a good indicator for students that a term denotes a *process*. Of the 185 process terms in the 807-term list, about half of the terms (99, or ~54%) had a *–tion* (or *–sion)* suffix. However, a term such as population can represent either a *thing* or a *process*, depending on context or use. And some terms such as warning coloration are solely *things*. The suffixes *–esis, –osis,* and *–ism* are found on ~ 15% (28) of the terms and the others have an assortment of endings. Again, the suffix clues are not always reliable, for example, since an hypothesis is a *thing*, not a *process*.

What makes processes different are their requisite conditions as to location, agents, inputs, substrates, transformations, outputs, chronologies, and functions. Learning a biological *process* is quite different from (and generally more complicated than) learning the nature of a biological *thing*.

Processes constitute a subset of propositional knowledge that poses problems in teaching, learning, and representation. More than *things* (objects with properties, and states), *processes* frequently are the delimiters of subfields in biology and they lend cohesiveness to research communities (Fisher, personal communication, November, 1997). Often they comprise the most difficult ideas to learn and understand, and pose the greatest barriers to communication across the subfields of biology.

Note that the *sequence of events* (time) idea is at the semantic center of a process. Interestingly, if you look at geographic place names on road maps and sea charts, *associated events* are sometimes used in naming places (e.g., Fort Defiance, Ten Sleep Canyon, Death Valley, Discovery Bay, Slippery Rock, and Cape Fear). Perusal of the aforementioned term list shows that this also occurs in biology. Terms such as *pacemaker, intertidal zone, salivary gland, seminiferous tubules, taste buds, tropical rain forest, urinary bladder, regulator gene,* and *photo receptor* are all process-based place names.

METAPHOR AND ANALOGY IN BIOLOGY LEARNING

Processes that are unfamiliar to us are often explained to us through the use of similes, metaphors, and analogies. When we tell students that the kidney is a filter, we are using a metaphor. When we expand that assertion into (+) and (−) feature mappings between the *target* (kidney) and the *analog* (filter), we are crossing the divide between metaphor and analogy. The kidney is like a coffee pot filter in that it separates substances found within a fluid, in part, on the basis of their particle size. The kidney is NOT like a coffee pot filter, for example, in its mechanism (glomerular filtration) and its ability to selectively reabsorb materials needed by the body. Think of an explicit, strength-of-relationship continuum from *simile* (weaker pole) to *metaphor* (the middle) to *analogy* (stronger pole). French (1995) claims that much of our learning depends on bootstrapping by analogy from our prior knowledge to the new content to be learned. Marshall McLuhan (Gordon, 1997, p. 71) once said, in deep parody of poet Robert Browning's famous words, "A man's reach must exceed his grasp, or what's a metaphor?"

THE NATURE OF NON-LITERAL LANGUAGE

Science is often perceived as being predicated on confidence that its language is precise and unambiguous − or, literal (Ortony, 1979). Would that this were so! Histories of science show us that we often expand our scientific knowledge by using metaphor and analogy, and that "language, perception, and knowledge are inextricably interdependent" (Ortony, 1979, p. 1).

Before the advent of constructivism in science education, metaphors and analogies were seen as wobbly instructional crutches useful in teaching the less able student. They were considered more poetic and fuzzy than they were scientific and rigorous. The better student (said to be at a "formal operations" level by Piagetian stage theorists) was thought to be cognitively fit enough to move forward in understanding without the aid of such pathetic intellectual prostheses as metaphors and analogies.

However, if the *search for similarity* inherent in creating and using metaphors and analogies is a fundamental cognitive pathway as Lakoff (1987) and others claim − and one that can lead to deep, not just shallow, understanding, then prior knowledge is further buttressed as a principal limiting factor in learning.

Then too, *the metaphor of construction* is an apt one that reflects not only how organisms come to know their environments, but also how science students come to know biology. If the goal of biology instruction is to place important, well-integrated, generative knowledge into long-term memory, then metaphor and analogy appear to be appropriate and powerful semantic tools.

Claus Emmeche observes:

> Biologists have often employed a range of metaphors to describe the real nature of organisms, and the metaphors have typically been borrowed from the technology that happened to be most fashionable at the moment. An ant, for example, can be viewed as a mechanical piece of clockwork, with precise, finely tuned parts, each with its distinct function. From a subsequent perspective, the ant can be viewed as a piece of energy technology: a thermodynamic design that – in analogy to a steam engine – consumes chemically bound energy by combustion and performs work while developing heat. Today we might view the ant as a little computer with associated sensory and motor organs: it processes a mass of information about the external world and reacts by feeding back various responses. (AAAS, 1994, p. 1901)

E. O. Wilson (AAAS, 1994, p. 1901), the famous Harvard and entomologist and evolutionary biologist, sees metaphors as vital to biological thought and writes, "Much of the history of biology can be expressed metaphorically as a dynamic tension between unit and aggregate, between reduction and holism. An equilibrium in this tension is neither possible or desirable. As large patterns emerge, ambitious hard-science reductionists set out to dissolve them with nonconforming new data. Conversely, whenever empirical researchers discover enough new nonconforming phenomena to create chaos, synthesizers move in to restore order. In tandem the two kinds of endeavors nudge the discipline forward.

POPULAR BIOLOGY WRITERS: AN ANALYSIS OF THEIR USE OF METAPHOR AND ANALOGY

Many who enjoy learning biology have read the popular press publications of Dawkins, Gould, Mayr, Thomas, and Wilson. They are masters at making biology exciting and interesting. Each has a unique writing style and uses metaphor and analogy effectively. Hackney and Wandersee (in press) have investigated representative writings of each author. Those who seek to teach biology ought to take note of the approaches these successful popular authors employ. While all of the authors in that study were male, we would now include female authors such as Diane Ackerman, whose works are growing in popular appeal.

While all five of the authors we studied used metaphor and irony as literary devices, we found small but important differences among them:

1. Entomologist and evolutionary biologist E. O. Wilson and physician Lewis Thomas share a worldview that emphasizes the interdependence and kinship of life.
2. Stephen J. Gould and Ernst Mayr highlight life's unity, diversity and complexity, evolution, and the history and nature of biology.
3. Dawkins tackles the big issues in evolutionary theory.

After carefully analyzing representative books by each of our five acclaimed, popular biology authors, we (Hackney & Wandersee, in press) found that they drew on a total of 23 different domains, from *architecture* to *unions*, in generating analogies. The total number of domains employed by each author was as follows: Dawkins – 20, Gould – 22, Mayr – 20, Thomas – 20, and Wilson – 17. This suggests that the expert biology "popularizers" rely on a diversity of subject areas for their analogical explanations – and similarly, biology teachers may wish to consider expanding their verbal horizons when crafting the best analogy for teaching a difficult biological concept.

Subject areas used as analogs most frequently by our five popular biology authors (when pooled) included: biology (n = 109), creative arts (n = 71), machines & industry (n = 62), military/conflict (n = 56), sports and games (n = 50), & architecture (n = 50).

We came away from our study realizing that the use of analogical explanations is quite a complex enterprise. Dawkins uses analogy as follows: "There is a tendency for metabolic rate to depend on body size in mammals generally. Smaller animals tend to have higher metabolic rates, just as the engines of small cars tend to turn over at a higher rate than those of larger cars" (1986, p. 106).

Like Gould, we found that Dawkins is careful to explain his analog, careful to explicitly state the point at which the correspondence between the analog and the target breaks down, and he often makes extended use of the same analog, rather than jumping from one to another. In short, Dawkins is a painter who stays with his chosen brush until the entire design is clear. (Did you notice the analogy I used to explain how Dawkins uses analogy?)

Each of the authors in our literary pantheon of popular biology is a powerful communicator. He recognizes the influence of metaphors from biology's distant past – but he also sees their inherent weaknesses. Wilson is particularly sensitive to the affective impact of his metaphors and is careful not to leave his readers with a sense of hopelessness. Gould teaches us that metaphors may be humorous – and thus easier to remember. His sense of humor is nowhere more evident than when he refers to the migratory movements of some animals as "no more peculiar than the annual winter migration to Florida of large mammals inside metallic birds" (1986, p. 30).

The bottom line I see is this: In purchasing popular biology books, people vote with their wallets. When they continue to purchase works by a given author, they are validating his or her written explanations and their effectiveness. It would be foolish for those who wish to become master biology teachers to ignore these data.

CHILDREN'S USE OF METAPHOR AND IRONY

Psychologist Ellen Winner (1988) has written an insightful book called *The Point of Words: Children's Use of Metaphor & Irony,* in which she points out that metaphor and irony are both forms of nonliteral language – and thus more daunting to understand than literal language. She distinguishes between metaphor and irony by showing that metaphor illuminates attributes of things in the world (and is thus a window on the learner's classification skills, whereas irony reveals the ironist's attitude about the world (and is a window on the learner's ability to ascribe intention and belief to others).

Winner (1988) points out that meaning is a slippery thing in human communications. Words are often not meant (e.g., "Turn on the air conditioning. I'm on fire!"); meanings are harder to derive from written than face-to-face communication (e.g., "I *like* your sense of design and balance." [said sarcastically and with grimace by the speaker]); and statements can be contrary to fact and yet true at some level (e.g., "All life comes from preexisting life"). The contradiction here, of course, is how did life begin?

Metaphor has the following characteristics (Winner, 1988): (a) it highlights certain attributes of an object or event; (b) it functions to clarify, illuminate, explain; (c) it can convey new information about objects and events; (d) it can reshape our thought (e.g., the universe as mechanical clock); (e) it reflects our conceptual framework; (f) it has *similarity* at its heart; (g) it poses a logic-analytic decoding task; and (h) it is seldom taken literally.

In contrast, Winner (1988) says, irony has these salient features: (a) it has *opposition* as its heart; (b) it poses a social-analytic task; (c) it requires detection of nonliteral intent, which is hard for children to do and so they sometimes take irony literally; (d) it has had less research study than metaphor; (e) its purpose is to critically comment upon an object or event; (f) it reveals the speaker's or writer's attitudes and thus has the effect of polarizing the audience; and (g) it has only a secondary purpose of description.

Interesting problems that face cognitive scientists studying metaphor include:

1. How can we account for our capacity to understand new metaphors?
2. How do metaphors reshape our mode of categorization?
3. How do metaphors become overused, die, and lose their metaphoricity?
4. Why are metaphors primarily juxtapositions of nouns?
5. Why did Aristotle claim that it is through metaphor "that we can best get hold of something fresh?"

Metaphor and irony are filters of objects and events – both can occur at the level of the sentence or pervade an entire text, but irony almost always has a victim, whereas metaphor rarely does (Winner, 1988). Irony requires a language-mature audience and flouts the traditions of ordinary conversation because it uses evaluative incongruity, contradiction, and incompatibility.

Piaget's research suggests that ability to understand metaphors is one of the last language skills to develop in children and usually occurs after ages 8 or 9. Typically, young children and novices focus on physical similarity rather than functional (relational) similarity. Although children spontaneously use metaphors from early on, they gradually stop using them and develop a preference for explanatory analogies. Children find irony much more difficult to detect and understand – instead interpreting it a lies or teasing. Irony also constitutes a much higher memory load. Interestingly, research shows no relation between a child's ability to understand irony and his/her ability to understand metaphor and analogy.

A key feature of metaphor and analogy is that they offer convenient ways to enter a new domain. We can reason analogically from a familiar domain to a new one. Both are nonliteral forms of communication and both require going beyond superficial appearances. The required ability to ascribe intentionality and belief makes ironic statements more difficult to interpret than metaphorical or analogical statements. Thus, one way biology teachers inadvertently mislead their students is when they use these devices without realizing some students will misinterpret the meaning they were to convey.

A RESEARCH STUDY ON THE USE OF ANALOGY IN SCIENCE TEACHING

Most biology teachers use analogies as an occasional instructional strategy, but researcher Thomas Mastrilli (1997) wondered just how often they are used and what forms they take. He also was interested in the reported basis for biology teachers' use of analogies in the biology classroom. In his use of the term analogy, Mastrilli included similes and metaphors, as well as visual comparisons. He studied, in ethnographic fashion, eight experienced, inservice, suburban and urban biology teachers considered to be good teachers. Teachers were not informed about the analogical focus of the study so as not to alter their normal teaching behaviors.

A total of 151 analogies were tabulated for the 40 biology class periods observed (mean = 3.8 analogies per period). Mastrilli (1997) found that the analogies his teachers used during the study fell into five categories: (a) simple/descriptive analogy (e.g., "like beads on a string"); (b) compound analogy (rasp/sandpaper); (c) spontaneous analogy ("like Jennifer's cloak"); (d) example analogy ("most common example are [sic] planaria"); and (e) visual analogy ("teacher twisted a plastic ladder to represent a DNA double helix").

Surprisingly, the biology teachers in Mastrilli's (1997) study tended to ignore their biology textbooks' analogies – preferring their own spontaneous ones or those drawn from their pool of previously spontaneous ones. Also, it should be noted that relatively few analogies were present in the biology books they used. The research characterized the biology teachers' use of instructional analogies as intuitive, superficial, and infrequent. He recommended that both new and seasoned teachers receive training about the appropriate use of analogy teaching models developed by Glynn (1989) and by Zeitoun (1984).

INSIGHTS GLEANED FROM THE MILLER ANALOGIES TEST (MAT)

The Miller Analogies Test has been used for years by some universities as a graduate school admissions standard. The 50-minute test consists of 100 multiple-choice analogy questions arrayed in order from simple to difficult. It is administered by The Psychological Corporation of San Antonio, Texas; an individual's MAT score (one's percentage correct) is considered valid for up to 5 years after the test was taken.

The Miller Analogies Test involves comparison of the relationship inherent in a given pair of words to a parallel relationship between another pair of words – of which, one word is missing. The words and relationships are drawn from general knowledge, natural sciences, social sciences, mathematics, literature, fine arts, grammar/linguistics/word play, and combinations of these subject areas. The examinee is asked to select, from four word choices, the word that most accurately completes the parallelism. For example, PASTEURIZATION is to POLARIZATION as KILLS is to ___ (a. repeats, b. separates, c. chills, d. classifies) (Spence, 1992, p. 129). In the interest of brevity, on the MAT it would be presented as PASTEURIZATION:POLARIZATION::KILLS: __ (a. repeats, b. separates, c. chills, d. classifies). One has to know that *pasteurization* is the controlled heating of milk, which has effect of the *killing* the microorganisms living within it. Similarly, *polarization*, in chemistry, has the effect of *separating* particles by charge.

It is this focus on analogical relationships that is so relevant to mapping biology knowledge. The pair relationships included on the MAT may involve: part/whole; part/part; member/group; degree and sequence; synonym/synonym; alternative names; antonym/antonym; intensity; chronology; size; spatiality; cause/effect; actor/action; actor/object; action/object; actor/function, action/function; mathematical; grammar/ linguistics/word play; event/place, creator/work-created; discoverer/discovery; individual/place of work; field of study/that which is studied; person or thing/associated event; person or thing/associated characteristic or quality; positive cause/effect; negative cause/effect; positive actor/object; and negative actor/object (Spence, 1992, pp. 18–29). Not only are such relationships important to the elaboration of knowledge on biology maps of a topic, but also, I have noticed, in construction of biology test items to probe students' understanding of that topic. A major European study of professional success in science recently found that analogical items were good predictors of such success. I am not surprised.

RESEARCHERS ATTEMPT TO UNDERSTAND WHY IQ TEST SCORES ARE RISING WORLDWIDE

James R. Flynn, a New Zealand political scientist who is based at the University of Otago, has noticed that average IQ-test scores throughout the world have risen steadily and markedly from the year 1918 on, but that this rise has been masked by the score-scaling system (Shea, 1996). While his best evidence comes from required IQ tests given to Belgian, Israeli, Norwegian, and Dutch soldiers, data from the U.S. and other countries also confirm this.

The so-called "Flynn effect" suggests that *environment* has a strong influence on what IQ tests measure. Since the biggest improvements have come on the sections of the tests that make use of pattern-completion and maze items, psychologist Patricia M. Greenfield hypothesizes that, increasingly, we live in cultures that reward visual-spatial thinking, and that people in contemporary societies need to, and do, develop intelligence in those areas (Shea, 1996). She claims that our new, visually intense world is our *eco-cultural niche* and, more and more, we thrive in visually rich environments. In this visual-cognitive milieu, biology knowledge maps may assume great importance as biology learning tools – combining language, space, and structure. As Manhattan cartographers Danniel and Jackson Maio see it, maps are captured freeze-frames of a changing subject area – providing information for day-to-day living and a lasting record that can benefit others (Dunlap, 1997, p. A24).

ON AMBIGUITY – FROM BABY TALK TO FATAL WORDS

Although, as our opening vignette indicated, we carefully enunciate for and speak in complete sentences to infants, we tend to leave out important contextual and aural cues when we communicate with older children and adults, especially if we consider what we are saying to be routine. In the classroom, such underspecification and ambiguity of language can lead to critical errors or gaps in understanding; in the real world, it can even lead to physical injury and death (e.g., air traffic control messages). In fact, it might be valuable to consider what can and does go wrong in such routinized but high stakes communication, so that we might better understand what can and does go wrong when we are communicating biology within a classroom setting or what can go wrong when mapping biology knowledge.

Steven Cushing (1994) has written a fascinating book entitled: *Fatal Words: Communication Clashes and Aircraft Crashes*. It looks at the role of language in aviation safety – using voice-mediated, air-ground communications tapes and transcripts as data. Its focus is on the confusions, omissions, and misunderstandings embedded in natural language that led to major aviation disasters. Similar interferences can be found in biology classrooms, as students assign different meanings than those intended by the professor, mis-hear certain words, and so on. For these reasons, this analysis seems relevant here.

Language problems detected in air traffic control messages fell into these categories: (a) structural ambiguities; (b) lexical ambiguities; (c) lapses into everyday speech; (d) homophony (words that sound alike but differ) and (e) speech acts. The language problems identified were often linked with other kinds of problems (Cushing, 1994).

For example, *problems of reference* were also apparent to the researchers. These were categorized as: (a) uncertain reference; (b) pronoun indeterminancy; (c) indefinite nouns; (d) "hear–back" problems; (e) mike lag; and (f) unclear hand-offs (Cushing, 1994).

In addition, *problems of inference* were found. These included: (a) implicit inference; (b) syntactic misdirection; (c) optional omission of the relative phrase "that is;" (d) lexical inference; (e) misconstruing statements of possibility for ones of permission; (f) use of indefinite verbs (e.g., expect, anticipate); (g) unfamiliar terms; (h) false assumptions; and (i) wishful thinking (Cushing, 1994).

There were also *problems of compliance*, which were divided into the following categories: (a) distractions and fatigue; (b) impatience (c) obstinacy and non-cooperation; (d) crew conflict or frivolity; and (e) overt rudeness (Cushing, 1994). Most instructors have seen all of these attitudes among their students at one time or another.

Finally, the last class of problems of possible relevance to biology teaching was sending *and receiving problems,* categorized as (a) message not sent; (b) message sent but not heard; (c) message sent and heard, but not understood; and (d) message sent, but forgotten (Cushing, 1994).

If you have a background in logic, linguistics, and aviation, you will undoubtedly understand all of the aforementioned categories. If not, do not despair. Some were undoubtedly generative for you immediately upon reading them. The ones I deem most relevant to biology teaching from each category are as follows:

1. lexical ambiguities due to imprecise speech;
2. "hear-back" problems due to lack of sensitivity to student feedback about what is being taught;

3. use of unfamiliar terms that nullify the effects of the teacher's intended precision of instruction;
4. the interference of distraction and fatigue, due to contemporary students' diminished attention spans; and
5. teaching messages intended to be sent and thought to have been sent, but not actually sent to the learners.

Please consider that if research revealed there are this many categories of problems in air voice-mediated air-traffic control communications (which are rather constrained in scope), it seems reasonable to expect to find even more problem areas and categories in biology classroom communication.

In the domain of aviation communications, the following "immediate fixes" were recommended after all the data were analyzed:

1. Controller's instructions should be read back rather than the pilot just replying "O.K."
2. Pilots should ask clarifying questions whenever they are puzzled.
3. Controllers should always give the labels (e.g, heading) along with the numbers they give out
4. Awareness is the first key to safety.
5. Controllers should be aware of their propensity not to listen carefully to "read–backs" from pilots (Cushing, 1994). It does not take much interpretation to translate each of these basic recommendations into parallel implications for biology teaching.

The researchers also extracted a set *of common concepts* from the voice-communication-induced incidents (listed here) and conducted concept-specific analyses as well. Here are the problematic concepts that were ultimately identified (Cushing, 1994):

1. maintain/cruise/ascend/descend
2. location & facility names
3. route
4. aircraft type
5. altitude/altimeter
6. time
7. direction/heading
8. speed/accelerate/decelerate
9. frequency
10. altimeter
11. weather level
12. relative movement (parallel, crossing L to R, closing, diverging)

We think that future research studies using the SemNet® semantic networking software (see Chapters 1 and 9), audiotapes and videotapes of biology lessons, transcripts of teacher–student interactions, and salient biology concept lists such as the one included at the end of this chapter, will allow significant progress to be made in understanding how to craft lesson scripts that optimize the words and sentences we use in biology teaching. Scripted lessons can be quite helpful when a teacher is presenting a subject for the first time.

SUMMARY

Meaning-making is a priority in language and communication at all ages. Scientists increase their ability to communicate precisely by developing a jargon unique to their discipline, but this gain in precise communication among themselves is counterbalanced by a reduction in the scientists' abilities to communicate with nonspecialists.

Since dynamic processes are generally more difficult to describe and comprehend than things, similes, metaphors and analogies are often used as aids in describing them. Metaphorical thinking about complex ideas is believed to occur automatically at the subconscious level as well as being used as a tool that is invoked consciously. Popular science writers make extensive use of metaphor and analogy. A key feature of metaphor and analogy is that they offer convenient ways to enter a new domain. Irony is often used in language as well but is less effective in promoting clear communication, especially with children.

In the classroom, underspecification and ambiguity of language can lead to critical errors or gaps in understanding. In the real world, ambiguity can lead to physical injury and death (as in misunderstood air traffic control messages). Much more research is needed to allow significant progress to be made in understanding how to craft lesson scripts that optimize the words and sentences we use in biology teaching.

JAMES H. WANDERSEE

CHAPTER 7

Using Concept Circle Diagramming as a Knowledge Mapping Tool

How Lana is Learning Biology

Everyone gathered around Lana Preszler's desk. No other first-grader at Natoma Station Elementary School had ever seen an insect that looked like this one. Lana had collected it (with her father's assistance) from the freshly painted exterior wall of her family's home in Folsom, California. She had kept it in her "bug farm" container so she could bring it to school to show to her classmates, and now none of them could tell her what it was. No one else at her school, neither the teachers nor the parents, recognized it either!

However, Lana didn't give up her quest to identify her find. Instead, she went to the family computer, and, with her dad's help, performed an Internet search. The search terms ("black white insect") resulted in many "hits," but a photo displayed on a Cornell University web site (based in Ithaca, New York) caught Lana's eye. It was an insect that appeared to be similar to hers, and it was labeled as an Asian longhorned beetle.

So, she and her father sent an E-mail message to E. Richard Hoebeke, the Cornell scientist who had established that web site, telling him about her unknown organism. In reply, Hoebeke asked her to send him the insect at once, and then explained how she should go about shipping it to him. As a result, the insect arrived alive and well at Cornell University. It was a better ("more pristine" in Hoebeke's words) specimen than any of the others of its species in Cornell's 5-million-insect collection. Hoebeke knew at a glance that this was not the maple-tree-destroying Asian longhorned beetle currently troubling New York residents, but a banded alder beetle (see Figure 7.1).

He asked Lana if he could have it for the Cornell insect collection and Lana said, "yes." In thanking her, Hoebeke also recommended she donate additional specimens to several universities in her home state, specifically UC—Davis and UC—Berkeley.

Interestingly, he added, this harmless beetle, classified as Rosalia funebris, is known to be attracted to wet paint! In fact, a 1995 scientific note published in the Pan-Pacific Entomologist by a University of California—Berkeley entomologist, E. Gorton Linsley, stated that he had observed this, and then he speculated that a volatile paint chemical may mimic an attractant pheromone used by these beetles prior to reproduction.

Figure 7.1. The banded alder beetle; photo by Frank DiMeo/Cornell University Photography.
Photo courtesy of The Cornell Chronicle.

And that's why a major research university in New York state has a notable preserved
specimen of a banded alder beetle, collected on May 18, 1997, by a curious 7-year-old
California school girl named Lana Preszler, to be found in Drawer 20 of its
Cerambycidae cabinet for scientists from all over the world to study (adapted from
Friedlander, 1997).

THE PROBLEMS A BEGINNING LEARNER FACES

It seems obvious from the opening vignette that young learners like Lana <u>are</u>
interested in the natural world and that life science instruction can help them to make
sense of their out-of-school experiences. Lana's expanding entomological knowledge
base can be seen in her transition from the all-encompassing children's term "bug"
(not the entomologist's *bug*) to *insect* to *beetle* to *species of beetle*. It is a real-life tale
of concept differentiation involving, among other things, the elucidation of inclusive-
exclusive relationships.

All of us think with concepts. Humans use concepts to classify and explain
objects and events. Biology has its own specialized concepts. Once concepts are
understood, relationships between them can be grasped and knowledge structures can
be constructed by the learner. For the purposes of this chapter, a concept can be
defined as a regularity in objects, events, or properties which has been given a name –
such as *insect*, *reproduction*, or *white*.

Concept development poses a challenge to beginning learners – whatever their
chronological age. Until they have discovered the meaning of a basic set of
foundational concepts in the area of interest, they cannot begin to apply this
knowledge to understand more sophisticated concepts and relationships. The most
difficult (and slowest) phase of mastering a new subject area is getting the
fundamental concepts straight. Unless teachers recognize this and adjust the
instructional pace accordingly, learners can get lost early in the course, lose interest
fast. and depend upon rote learning to mimic actual understanding.

A METACOGNITIVE TOOL FOR LEARNERS LIKE LANA: OVERVIEW

Metacognitive (reflective thinking) tools have the potential to help each of us mark
our path and prevent us from getting lost as we learn new concepts and relationships.
A *concept circle diagram* (CCD) is a metacognitive tool specifically designed to help

beginning learners navigate unfamiliar conceptual waters. Specifically, it is a form of graphic representation used to depict inclusive-exclusive relationships among or between bounded, taxonomic concepts (Wandersee, 1987). Markman (1989, p. 14) points out that nested, class-inclusion relations are "... a pervasive and extremely important kind of organization of categories." Because a concept circle diagram involves clusters of five concepts or less, it is, by design, less complex than a concept map. It uses the metaphor of apportioning and "fencing-off" conceptual space.

Science educators Alfred Collette and Eugene Chiappetta (1994) observe that "Wandersee has extended the work of Novak and Gowin, providing [middle and high school] science teachers with more explanations and techniques for graphically representing scientific knowledge" (p. 68).

Psychologist Paul Hettich (1992) writes, in his popular study skills manual for college students:

> The major advantages of the concept circle diagram are ease of construction and visual effectiveness, especially when you wish to represent a small number of concepts in graphic form. To illustrate, you could remember the categories and subcategories of seed plants [angiosperms] for your botany class by writing them on a sheet of paper and then rehearsing them [or] you could construct...a concept circle diagram like Professor Wandersee devised for his students. (p. 205) ... When important information is not presented visually, try to make it visual (p. 208).

In a recent Association for Supervision and Curriculum Development (ASCD) publication, curriculum and professional development specialist David Hyerle (1996) presents his vision for integrating teaching, learning, and visual tools. He introduces, explains, and endorses concept circle diagrams as "task-specific organizers" for yielding "holistic images that students can easily grasp and mentally manipulate" when learning science (p. 60).

Before more of the theory behind concept circle diagrams is presented, it may be helpful to look at a simple (as opposed to compound) concept circle diagram (see Figure 7.2). This learner-constructed concept circle diagram represents learner-selected aspects of a newspaper-reported breakthrough in biological knowledge about what controls blood flow in the human body (Blakeslee, 1997). Recall that one aspect of the goal of making US citizens scientifically literate is helping them to interpret the science they see reported every day in the news media. The CCD diagram shown in Figure 7.2 represents a learner's step toward such self-actuated meaning-making.

Note that, in Figure 7.2, the learner has given his concept circle diagram a descriptive title, represented conceptual relationships spatially, and written a sentence summarizing the main idea of the diagram. The basis of his diagram was a *New York Times* article (Blakeslee, 1997) that reported the progress physiologists are making in understanding how the human body regulates the flow of blood in individual tissues – a basic science discovery with profound applied science implications for the treatment of everything from heart attacks to high blood pressure.

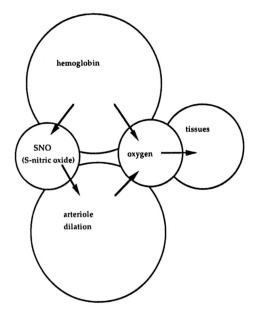

Figure 7.2. Blood as an auto-regulatory agent.
When a tissue needs oxygen, hemoglobin in the red blood cells in a nearby arteriole changes
shape, releasing both oxygen and a special form of nitric oxide. The nitric oxide causes the
arteriole to dilate, increasing the local blood flow.

Formerly, available evidence indicated that local blood flow changes are due primarily to muscle-controlled changes in diameter of artery walls, and that the blood itself is a relatively passive fluid. Today, researchers have reason to think that when a nearby tissue needs oxygen, the blood's hemoglobin changes its shape – releasing some oxygen plus a special form of nitric oxide (SNO) that dilates the arteriole, increasing blood flow into the capillaries. The bottom line is: the blood itself is the controlling (auto-regulatory) agent – it's not as passive a fluid as scientists once thought! The learner attempted to encapsulate some key aspects of this article in his diagram. Experience using CCDs in the science classroom has shown that such a diagram (albeit "minimalist") not only reminds the constructor of the whole article and the topic's most salient ideas, but also *visually triggers* the learner's recall of many of the unrepresented supporting details. It is this *visual distillation effect* that gives value to self-constructed graphics like the concept circle diagram – otherwise we could simply use descriptive prose.

THE BASIC IDEAS BEHIND CONCEPT CIRCLE DIAGRAMS

Concept circles can be defined as two-dimensional, labeled geometric figures that are constructed by the knower to be isomorphic with his/her personal understanding of the conceptual structure of a small, manageable cluster of concepts. When a title and an underlying explanatory sentence are added by the knower, this piece of "cognitive art" (Tufte, 1990) is known as a concept circle diagram (see Figure 7.3 for the format of a concept circle diagram). This author invented this tool and the technique for using it during the summer of 1984 while doing postdoctoral work in biology education at Cornell University under the direction of noted biology educator Joseph D. Novak.

The intent of this technique is to introduce beginning learners to concept-based learning and principles of metacognition by enabling them to represent small clusters of concepts (five or less) on paper, by applying simple rules for labeling, sizing, coloring, and positioning template- or computer-drawn circles. The concept circle diagram is the first in a series of increasingly powerful Ausubelian (now Human Constructivist) metacognitive tools to be used by science students – to ease students' transition to concept maps and vee diagrams.

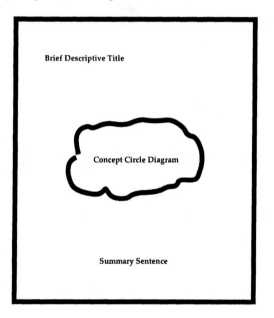

Figure 7.3. Concept circle diagram format.

The circle template used in the technique was developed (Wandersee, 1987) after a research study of mine yielded a set of circle sizes that appeared to my human subjects (undergraduate biology students) to be arranged in ascending order of enclosed circular area in a ratio of 1:2:3:4:5. Since it has long been recognized that human estimates of circular area deviate from actual (mathematically calculated) circular area, Concept circles are "psychologically sized" rather than "mathematically sized," using data regarding students' perceptions and calculations made with Stevens' Power Law (Stevens, 1975). That is, the circles were designed to *appear* to be 2, 3, 4, and 5 times larger (in enclosed circular area) than the unit circle. The resulting standard set of circles can thus be used by the learner without "visual dissonance" to represent relative importance, levels of hierarchy, quantity, or chronology.

Fifty templates for student use were constructed in a home workshop (by drilling the research-determined 1", 1 7/8", 2 1/8", 2 1/2" and 3 1/8" holes in template-size 7" x 5" pieces of 1/8" tempered hardboard. Subsequently, a commercial architectural template manufacturer produced thousands of brightly colored (orange), unbreakable, plastic concept circle templates that students today find helpful in making their diagrams look – to use their own words – "professional" (Wandersee, 1987). See Figure 7.4 for a photograph of the commercially manufactured concept circle drawing template.

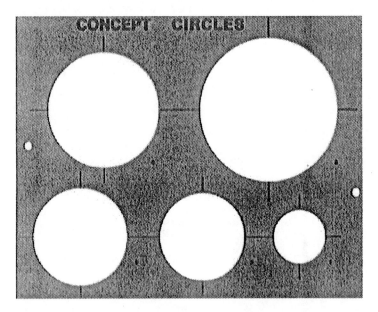

Figure 7.4. Concept circle drawing template.

The concept circle diagram technique itself is based on research findings in the fields of visual perception, the psychology of human memory, and was subsequently influenced by Novak's (1998) human constructivist learning theory. At first glance, concept circle diagrams often remind the viewer of the Venn diagrams they encountered as mathematics students. However, there are important theoretical differences. Venn diagrams involve no more than three circles, are used by logicians to represent syllogisms – not concepts, and employ special shading conventions to highlight logical intersections intended to lead the constructor to logically reasoned conclusions. In strong contrast, the starting point for concept circle diagrams was the relationship diagrams (called Euler's Circles) used by a mathematician, Leonhard

Euler (1707–1783) who preceded Venn (see Figure 7.5 for an illustration of Euler's Circles).

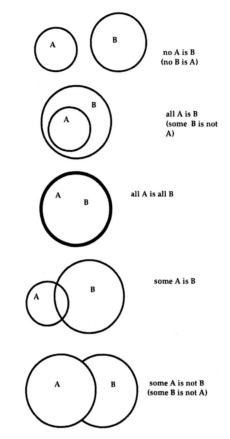

+Note: Darker line shows superimposition of circles.

Figure 7. 5. Examples of Euler's Circles.

In 1894, Venn pointed out that logicians borrowed the use of diagrams from mathematics during a time when there was no clear boundary line between the two fields. Line segments, triangles, circles, ellipses, and rectangles were all used to diagram categorical propositions during the early development of logic as a discipline (Wandersee, 1990, p. 927).

In his 1768 *Lettres a une princesses d'Allemagne*, mathematician Leonhard Euler used spatially positioned pairs of circles to depict five kinds of relationships. The origin of Euler's Circles can be traced back (in part) to the five cases of possible intersection of the famous Circle of Apollonius (262–190 B.C.). Martin Gardner (1968) suggests that Euler's Circles were eventually replaced by Venn's Diagrams because Venn's system fit Boolean class algebra so well. In my work, Euler's circles were adapted for use as a *meaningful learning theory*-based metacognitive tool. Concept circle diagrams go further than Euler in that they permit representation of a sixth relationship and up to five circles. In addition, graphic conventions were developed for labeling, coloring, depicting variables, and for telescoping one such diagram into another.

From the beginning, the assumption was made that explicit graphic conventions would allow the concept circle diagrams constructed by various persons to be easily decoded and assessed by any reader familiar with those conventions. Contemporary road maps, for example, always locate the compass point "north" at the top of the map, and we depend on this convention for quick orientation. Similarly, rather than being restrictive and confining, work with classroom science teachers suggests that such diagram conventions allow diagram constructors and their readers to share meaning more easily. Again, the value of a common structure and common conventions that are shared by practitioners is that it can be empowering, supporting easy communication and feedback.

"THE RULES" FOR CONSTRUCTING CONCEPT CIRCLE DIAGRAMS

The latest version of the graphic conventions (students call them "the rules") to be used in constructing a concept circle diagram is as follows.

1. A science concept is a pattern we see in nature (or need to invent in order to understand nature). Such patterns have accepted names or word labels (e.g., crystal, circulation, or eukaryotic). A single example of a science concept is called an instance of that concept (e.g., a barracuda is an instance of a marine fish).
2. Higher order concepts, composed of other concepts (e.g., (genotype, osmosis, density), are sometimes called constructs, and are treated the same way as concepts in concept circle diagrams.
3. The size of any concept circle represents its importance in the diagram. You can choose from five descending circle sizes.
4. Let a circle represent a particular science concept.
5. Print (not write) the name of that concept inside the circle using lowercase letters.
6. Within the diagram, only the first letters of proper nouns and adjectives (Hawaii, Bengal tiger) should be capitalized.
7. All concept labels should be centered horizontally and printed horizontally, with curved labels being permitted only when words are long.
8. When you want to show that one concept is included within another (fish are vertebrates), draw a smaller concept circle within the larger one and label it in the same way as described earlier. The larger circle then represents the more inclusive concept, the smaller one, the less inclusive concept.
9. Whenever you want to show that some instances of one concept are also included under another concept (e.g., some but not all of an organism's DNA is found in the nucleus of the eukaryotic cell), you can draw partially

overlapping circles and print the concept labels within the nonoverlapping areas of each.

10. When you want to emphasize that two concepts are NOT directly related (e.g., no fish are mammals), draw separate circles and label them.

11. When you want to show that the boundary of a single concept (e.g., life) or the boundary between two concepts (birth control and conception control) is not clearly understood, use a broken circle (a circle made of short line segments separated by equivalent-length spaces) to represent such "fuzzy" boundaries.

12. You may use up to five component circles in a single concept circle diagram. This limitation is (conservatively) based upon the 7+/– 2 rule for the processing capacity of short-term or working memory. These two to five circles can be separate, overlapping, included, or superimposed. If you find you have a need to represent more concepts than that, see the instructions for "telescoping" one diagram into another – as mentioned later in this rule set.

13. To show that two or more concepts are virtually equivalent when context is disregarded (e.g, hydrochloric acid and muriatic acid), use superimposed circles. Show that one circle lies directly on top of another by representing that superimposed set of circles with a single, thicker-lined circle. Then place the virtually equivalent concept labels inside it.

14. The relative sizes (bounded, inner areas) of the circles comprising your concept circle diagram can be used to represent relative levels of specificity for the concepts depicted (bigger circles standing for more general concepts, smaller for more specific ones).

15. Alternatively, the relative sizes of concept circles can be used to represent relative quantities or relative variable values for a given concept (e.g., number of known species, range size, biomass).

16. The plastic concept circle drawing template offers four larger circles that appear (to the human eye) to be 2x, 3x, 4x, and 5x the smallest circle (also known as the unit circle).

17. If quantity or value is the organizer of the diagram, a parenthetical lowercase "n" (n) should be positioned directly following the main concept label to show that a numerical value ("number") is implied.

18. Chronological relationships can be represented in a concept circle diagram by drawing an appropriate set of concentric circles with the smallest circle used to represent the oldest or starting concept to be depicted (e.g., spontaneous generation, protoplasm, prophase). If chronology is intended to be the organizer of the diagram, a parenthetical, lowercase "t" (t) should be positioned directly after the central concept label to indicate a "time" relationship.

19. A complete concept circle diagram consists of: (a) a descriptive title in the top, upper-left area, (b) a labeled and logically positioned set of concept circles in the middle, and (c) an explanatory (propositional summary) sentence placed towards the bottom of the page. Print, do not write, throughout the entire diagram. See Figures 7.6 and 7.7 for concept circles exemplifying some of the aforementioned rules.

Common Thermometers

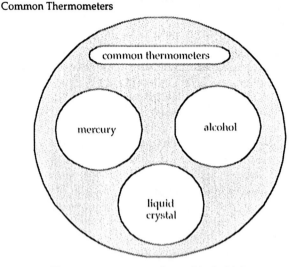

Thermometers commonly used in the biology
classroom include the mercury thermometer,
the alcohol thermometer, and the liquid crystal
thermometer.

Figure 7.6. Example: Common thermometers.

20. One concept circle diagram can be connected to another by a graphic
    convention called "telescoping." Broken straight lines are drawn tangentially
    to link the source concept circle to its related major concept circle in the
    "pulled-out" diagram that is located adjacent to it on the right. Via
    telescoping, a parade of diagrams (typically read sequentially from right to
    left) can be constructed by the learner (side-by-side on a scroll-like piece of
    paper) as instruction proceeds throughout the week, throughout the unit, etc.
21. Most concept circle diagrams can be improved by redrawing and revising
    them. It is rare that a person's initial diagram will accurately represent what
    he/she means.
22. On the final version of their concept circle diagrams, students may wish to use
    a brightly colored set of highlighter pens to color in parts of their diagram in
    order to make the relationships between concepts clearer. The following,
    perception-based rules then apply to such coloring.
23. A circle should be left uncolored if the concepts that it includes (within it) do
    not exhaust all the possibilities (e.g., if DNA were the only included concept
    within a circle labeled nucleus, the nucleus circle would not be colored in,
    because there is more than just DNA inside a true cell's nucleus). Conversely,
    under the topic of gases, if the included concepts were oxygen, nitrogen, and
    various other gases, and if the main circle were labeled air, the air circle would
    be colored in because that contextually exhausts the possibilities; the various

other gases concept could consequently be telescoped to include argon, carbon dioxide, water vapor , and trace gases.

24. The overlapping portion of any two concept circles should be colored in using alternating parallel strokes of both component circles coding colors.

25. Use colors that are as widely spaced on the visible light spectrum (referent acronym: ROYGBIV) as possible when coloring adjacent or included circles. This makes it easier for your eye to separate them when viewing.

Relative Emphasis in the High School Biology
General Biology Course

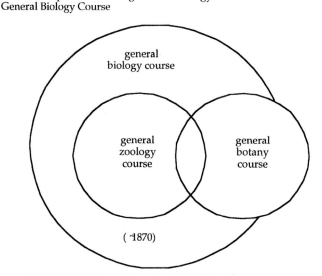

Nichols (1919) found that when high school botany and zoology courses were merged into a general biology course, plants were neglected, because most courses were taught by teachers educated as zoologists.

Figure 7.7. Example: Relative emphasis in biology course.

RESEARCH AND THE CONCEPT CIRCLE DIAGRAM

What makes the tool unique among the metacognitive learning tools currently available in biology education? Answer: It is easy to learn the rules for construction and to learn how to interpret other people's diagrams. Concept circle diagrams are not visually complex. In addition, they require only a pencil and paper to construct – although drawing and flow charting applications can be used for their creation and revision if computers are available. While detailed explanation and documentation of the tool's design is available elsewhere (Wandersee, 1987), here are a few examples to illustrate its research-based design:

1. The concept circle diagram's page design conforms to the identified eye-track pattern for reading the printed page in the Western world (see Figure 7.8 for an illustration of this tracking pattern).
2. The learner-generated printed title and propositional summary statement link the graphic to two different verbal indexes – consistent with Paivio's (1991) dual-coding theory.
3. The human eye's perceptual field has been determined to be roughly circular.
4. Evidence suggests that a circular figure is easier for the human visual cortex to process than more complex shapes.

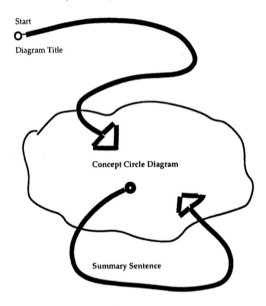

Figure 7.8. Concept circle diagram eye track pattern.

5. Lowercase printing is used throughout the concept circle diagram because visual perception research has shown that it is easier to decode such printing than it is to decode uppercase printing or handwriting.
6. Color-coding the CCD is an option available, based on Reynolds and Simmonds' (1981) finding that students distinctly prefer to view colored over black-and-white illustrations.
7. Research on edge detection suggests that spatially adjacent colors are easier to separate when their wavelengths differ significantly on the visible light spectrum – hence the coloring rule to use spectrum-spaced colors for coloring adjacent or included circles.
8. The maximum number of colors (five) that is permitted in a single concept circle diagram lies well within the limits of human color memory's capacity.

Human color memory allows us to remember (not detect) only about 24 saturated hues (shades) of color (Levy, 1987).

9. Concept circle diagrams employ direct labeling. This eliminates the need for *double scanning*, switching one's gaze back-and-forth from legend to graphic – a known cause of graphic inefficiency.

10. The circle sizes to be used in concept representation were experimentally chosen to match humans' skewed perception of circular area.

A CHECKLIST

The following checklist items (Wandersee, 1987) can be helpful for reviewing and evaluating a CCD. They are grouped by characteristics, as (a) a basis for comparing the quality of the biological representations of a single student across topics, and, (b) a basis for comparing the quality of one student's representation of a cluster of concepts on a single topic to those of the other students by tabulating weighted points (Table 7.1).

Table 7.1. A Checklist for Evaluating Concept Circle Diagrams

---

Basic Communication (COM)
- Is the student's concept circle diagram legible and interpretable? (If the instructor judges it is not, have the student orally explain the parts of his/her CCD on audiotape and submit the accompanying tape with the diagram).

Demonstration of Scientific Understanding (UNDR)
- Are concepts selected for representation central and/or relevant to the topic?
- Are the concepts presented in a scientifically valid way to show exclusive-inclusive relationships between concepts?

Quality of Dual Coding (DUCO)
- Does the descriptive title the student composed fit the diagram?
- Does the explanatory sentence the student composed fit the diagram?

Effective Use of Graphic Options (OPT)
- Was color used appropriately to clarify the meaning of the diagram?
- Were time-based or number-based concept circles used appropriately to clarify the meaning of the diagram?
- Was telescoping of concept circles used appropriately to clarify the meaning of the diagram?

Application of Graphic Conventions (RUL)
- Did the student follow the rules for making concept circle diagrams?

EVALUATION OF LEARNERS' CONCEPT CIRCLE DIAGRAMS

Concept circle diagrams are an alternative means of evaluating your students' understanding of a small cluster of concepts involving exclusive-inclusive taxonomic relationships. The results can be used formatively in the midst of a unit (to monitor progress and diagnose learning difficulties) or summatively (to provide a *cognitive snapshot* of your pupils' understanding near the endpoint of instruction). It can also be used metacognitively by the learner, to probe the adequacy and depth of one's current understanding via the process of transforming it into a CCD. Such skills take time to teach but are well worth the effort.

EVALUATING UNDERSTANDING OF BIOLOGICAL CONCEPTS

The first criterion on the checklist determines whether or not the biology student's CCD can enter the evaluation process by itself, or if it requires an accompanying audiotaped explanation recorded by the student (in a carrel near the back of the classroom). It is surprising how many students' diagram legibility and time-on-task improves when this provision is in place. The provision eliminates the need for undue interpretation and excessive speculation by the evaluator.

When grading (instead of diagnosing – as in formative evaluation) concept circle diagrams, points can be assigned by item cluster (refer to previously introduced checklist). No points are allotted for the *Basic Communication (COM)* criterion because it constitutes the evaluation gateway. The remaining categories are decreasingly weighted in point value in the order listed, with valid *Scientific Understanding (UNDR)* always receiving the most points. The number of points per category is adjusted during the course of the semester to focus students' attention on different aspects of the CCD as they work toward proficiency in using the technique. Students are always given a copy of the point scale in use at the time – prior to their diagramming task.

SCORING AND GRADING CONCEPT CIRCLE DIAGRAMS

In my own scoring system, no CCD is worth more than 10 points. Once you are familiar with your own point scale, evaluation time may be as little as 4 minutes per diagram. If you teach multiple sections of a course, you can stagger assignments so you never have more than one section's diagrams to grade on a given day. This prevents evaluator fatigue and insures high quality feedback.

A minicopy of the checklist is stapled to each student's CCD prior to evaluation and grading, and instructor feedback consists of the tabulated total point score for the CCD and the instructor's handprinted comments in the form of occasional words and phrases placed near the relevant checklist item on the form. Such comments can explain the loss of points on a particular checklist item or praise aspects of the student's CCD related to that item.

In conducting biology education research, a weighting scale can be used to assign point value to each category. This scale can sometimes differ dramatically from the ones used to grade biology students' work, in order to tailor it to the study's research questions. Rather than set forth a single way of evaluating concept circle diagrams, mindfulness and multiple perspectives are encouraged (Langer, 1997).

Nobles (1993) used a three-category rubric for evaluating students' concept circle diagrams in her Ph.D. dissertation research. Her study examined how using concept circle diagramming affects middle school students' science concept learning. The categories were conceptual sophistication, graphic complexity, and mastery of technique. Performance on each of the rubric categories was considered to be dichotomous. That is, conceptual sophistication was either explicit or lacking, graphic complexity was simple or complex, and mastery of technique was yes or no. Percentage agreement among science content experts using her rubric was 0.95.

Nichols (1993), in her study of students' understanding of metamorphosis, used a method for scoring concept circle diagrams that was adapted from concept mapping criteria employed by Wallace and Mintzes (1990). These were (a) every valid concept received one point, (b) each valid concept depicted in a proper scientific relationship with another concept received two points, and (c) each alternative conception was assigned one *negative* point. She noted that "It seems essential to use various methods to more adequately analyze student understanding of insect metamorphosis, since that understanding is an integration of diverse and related pieces of knowledge." In support of such a position, White and Gunstone (1992) consider "mode validity" to be improved with the use of a variety of knowledge probes.

This author wishes to underscore this conclusion, since concept circle diagrams are not intended as stand-alone evaluation tools. They do, however, reveal understandings that are not commonly activated by pencil-and-paper test items (Nobles, 1993). Larkin and Simon (1987, p. 65) confirm this when they write from an information processing perspective, "Diagrammatic representations . . . typically display information that is only implicit in sentential representations [written propositions] and that therefore has to be computed, sometimes at great cost, to make it explicit for use."

Every graphic metacognitive tool has characteristic features which must be considered during the selection process. Using Table 7.2, the reader can compare features of some graphic metacognitive tools – concept circle diagrams, concept maps, and SemNet® semantic networks. One tool is not clearly superior to another, just different. Knowing about the differences allows the biology educator to make informed choices.

Each of the graphic metacognitive tools in Table 7.2 can present a challenge to the learner. Each requires instruction in interpretation and construction. Some biology educators have viewed them as mere note-taking devices. Such a perspective ignores the opportunity to help the learner "learn how to learn" biology – to use Novak and Gowin's (1984) phrase. The learner needs to be made aware of the theory behind the tool to understand why it works and why rote-mode learning, although initially easy, is ultimately less effective and empowering. All of the aforementioned tools focus the learner on meaning-making and involve students in constructing their own knowledge base and taking charge of their own learning. It typically takes 1–2 months for a learner to master and become comfortable using a particular tool. Ultimately, the tool can become a prosthetic device, a memory extension useful in supporting the student's thinking.

Table 7.2. A comparison of some salient features of three graphic metacognitive tools.

Features	Concept circle diagrams	Concept maps	SemNet® semantic networks
Computer-independence	yes	yes	no
Visualizability of product	complete	complete	partial
Level of capture complexity	low	medium	high
Color coding	yes	yes	anticipated
Drawing template availability	yes	no	not applicable
Merging of sub diagrams	easy (telescoping)	difficult (cut & paste)	easy (merge nets)
Spatiality	2-D	2-D	n-D
Metaphor	fences	map	network, net
Memory load	low	medium	high
Explicitness of hierarchy	medium	high	low
Labeled links	indirect	yes	yes
Bi-directional links labeled	no	no	yes
Maximum number of concepts	5	~50	~32,000
Nuanced learning curve	gentle	moderate	steep
Analytic power	low	moderate	high
Evaluation of content validity	~5 min.	~10 min.	~20 min.

Fisher (1995, p. 12) holds that "...the process of constructing a network is much more valuable than the product." This is true to the extent that the product of a metacognitive tool is always just a "snapshot" or a "placeholder" of one's current knowledge – a known point of departure for further conceptual change. Psychologist Donald Norman would probably call each of these tools "affordances" that enable us to leverage our understanding of biology. Tools such as this one can really increase our "knowledge advantage." It reminds me of Archimedes' *still undisproven* claim when he finally understood how levers worked: "Give me a place to stand and I will move the earth!"

Throughout her lifetime, Lana will need to understand much more about the living world than the members of the generations that preceded her. Not only will she need to learn biology in her youth – she will need to learn how to build upon, update, and expand her biological knowledge throughout her life span. Metacognitive tools, such as concept circle diagrams, can help her do all of these things. Here is a concept circle diagram such as Lana might make to represent some of what she learned from her mystery beetle experience (Figure 7.9).

Lana's Mystery "Bug"

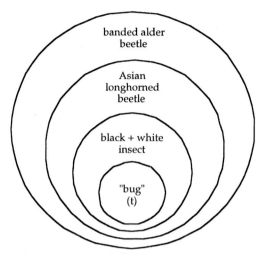

Lana found an unusual-looking "bug" in her yard, used
an Internet search engine with the term "black and white
insect," found a picture of an Asian longhorned beetle that
resembled her specimen, and sent an E-mail message to a
beetle expert who determined she had collected a banded
alder beetle.

Figure 7.9. Lana's mystery bug.

SUMMARY

All of us think with concepts, and biology has its own specialized concepts. Once
concepts are understood, relationships among them can be grasped and knowledge
structures can be constructed by the learner.

Metacognitive (reflective thinking) tools have the potential to help each of us
mark our path and prevent us from getting lost as we learn new concepts and
relationships. A concept circle diagram (CCD) is a metacognitive tool specifically
designed to help beginning learners navigate unfamiliar conceptual waters.
Specifically, it is a form of graphic representation used to depict inclusive-exclusive
relationships among or between bounded, taxonomic concepts. The major advantages
of concept circle diagrams are ease of construction and visual effectiveness,
especially when you wish to represent a small number of concepts in graphic form.
The origins of CCDs and rules for their construction have been described.

Concept circle diagrams are an alternative means of evaluating students'
understanding of a small cluster of concepts involving exclusive-inclusive taxonomic
relationships. The results can be used formatively, in the midst of a unit, to monitor
progress and diagnose learning difficulties, or summatively, to provide a cognitive

snapshot of pupils' understanding near the endpoint of instruction. It can also be used metacognitively by the learners, to probe the adequacy and depth of their current understanding via the process of transforming their current understanding into a CCD. Such skills take time to teach, but are well worth the effort. The teacher and learner need to be made aware of the theory behind the tool to understand the reasons for its rules, why it works, and why rote-mode, unmindful learning, although initially easy, is ultimately less effective and empowering.

JAMES H. WANDERSEE

CHAPTER 8

Using Concept Mapping as a Knowledge Mapping Tool

Nina Captures an Elusive Beast

An enthusiastic 19-year-old biology student named Nina was reading an article published in *Science*, the journal of the American Association for the Advancement of Science. She had previously been taught how to construct concept maps in her introductory college biology course. The article her teacher had asked her to read was entitled, "Entomologists wane as insects wax," by Constance Holden. The teacher had provided her with five "seed concepts" that were to appear on the final version of her concept map of the article – namely, *chemical prospecting, entomology, human problems, insects,* and *world biota survey*. Nina read the article, constructed an initial concept map, then checked her map against the article's contents by rereading them both. She kept editing and re-editing her concept map until she felt the gist of the article was now represented on paper. The entire task of reading and concept mapping the article took Nina about an hour – more than twice as long as she would have normally taken to read and study a 3-page scientific journal article such as that one. "As you peruse my concept map," she said, "remember that I have intentionally kept it rather small – sort of a mental skeleton which reminds me of the complete 'beast.' Having mapped it, I know it much better than if I had merely been asked to read it. It feels like I have captured its essence successfully, and it cannot escape. Aha!" (see Figure 8.1)

It is easy to relate to Nina's joyful feelings of *knowledge capture*. This author has been using concept maps for teaching biology and for research in biology education since 1980 (Mintzes, Wandersee, & Novak, 1997; Novak, 1998; Novak & Wandersee, 1990; Trowbridge & Wandersee, 1994, 1996, 1998; Wandersee, 1992a, 1992b; Wandersee, Mintzes, & Novak, 1994). During the 1980s, I spent eight summers in Ithaca, New York, working with Professor Joseph D. Novak, a professor of biology and professor of education at Cornell University, and the person who has been called the "father of concept mapping."

Together we have coedited a special issue of the *Journal of Research in Science Teaching* on concept mapping, cotaught high school teacher workshops, taught graduate-level science education courses, and produced articles, book chapters, and books on it. So, in order to insure a fresh treatment of the topic, I have organized this chapter in an unconventional way – using the organizational format of Frequently Asked Questions (FAQs).

127

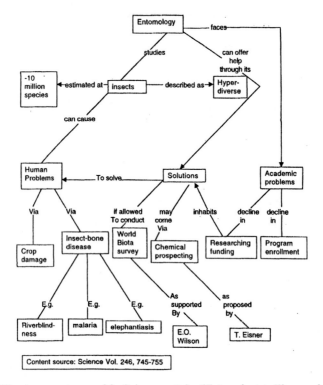

Figure 8.1. Nina's concept map of the Science *article, "Entomologists Wane as Insects Wax,"
by Constance Holden. How does this differ from a Novakian map?*

FAQ-1. WHAT IS A CONCEPT MAP?

It's not an intuitive thing to evaluate your own learning. In general, it is easier for both teachers and learners to evaluate "doing" rather than "thinking" (Griffard & Wandersee, 1998). Thus, metacognitive tools that capture and display an individual's knowledge support reflection and evaluation. Concept maps and other metacognitive tools are also central to the success of what Novak calls a *human constructivist* approach to learning science (Novak, 1998; Okebukola & Jegede, 1998).

Concept mapping is a form of graphic representation invented by science educator Joseph D. Novak and his graduate students at Cornell University (Stewart, Van Kirk, Rowell, 1979). Novak saw it as a vital part of a metalearning strategy (a way to help learners monitor their own learning). A concept map is a two-dimensional, tree-like, hierarchical array of circumscribed concepts linked together by lines that are labeled with linking words. It can be read by starting with the top (superordinate) concept and reading down the links and concepts of each branch of the map. Its hierarchical structure is intended to parallel the way the brain stores knowledge hierarchically (Wandersee, 1990).

Although there are other mapping strategies (e.g., mind mapping, concept webbing) in the literature that are generically labeled "concept mapping," most differ dramatically from the type that Novak invented, and are not theory-driven or intensely researched. Because Novak's concept map was the first to be widely used in science education, the term "concept map" is typically used by science educators to mean, specifically, Novak's type of concept map. It might also be noted that Novak has recently applied for a US trademark for the term "concept map" (Novak, 1998, p. iv).

FAQ-2. WHAT IS A CONCEPT?

Concepts are the ideas with which we think. Our brains are wired to search for patterns in *objects* or *events* or *properties* in our environment, and when such regularities are found, we use a label or a symbol to communicate with others about them. These labeled patterns are our concepts. For example, "rain" is the concept label we give to the *event* wherein atmospheric precipitation in the form of condensed, liquid water droplets falls from clouds to the earth's surface. The regularity in the concept of "pencil" is that of a cylindrical *object* made of a sharpened, baked, carbon-clay rod, typically surrounded by a cedar wood casing, and which serves as a hand tool that we can use to write or to draw images, and the concept, "density," is a *property* of matter that depends on both its volume and its mass (m/v).

In order to learn a concept meaningfully, we must perceive the underlying pattern or regularity to look for, and then we must practice applying it across a range of examples. Middle level noun concepts (e.g., tree) are typically learned first, and later, often during formal instruction, these are 1) subordinated, as in "tree is a seed plant;" 2) elaborated, as in "tree can be a gymnosperm or angiosperm;" and 3) differentiated, as in "gymnosperms may be pine, spruce, fir or ..."

One's understanding of a biology concept at a given point in time can be plotted on a continuum from highly rote to highly meaningful. A concept's status on this continuum can change for a given learner as an instructional unit progresses. Consider a student who learns by rote that spiders are placed in a category called *Class Arachnida*. (Her prior knowledge and experiences have led her to think spiders are a special type of insect and she tells herself that now she knows their "weird" scientific group name). Then she learns that the Greek word which is the origin of the category name *Arachnida* means "to spin" – a behavior she knows spiders exhibit when they make their webs. Then she learns that the same category also includes scorpions, mites, and ticks – with all members (including spiders) having four pairs of segmented legs and a two-section body plan comprised of a cephalothorax and an abdomen. This alerts her to the fact that spiders *cannot* be insects – they are fundamentally different. Later she learns the hierarchy of taxa that is used in binomial nomenclature and finds out that the designation, *class*, signals that this is a broad category, just below *phylum* (in this case, the phylum Arthropoda).

Can you see how the proposition that "Spiders belong to Class Arachnida." gradually acquires more meaning when biology instruction is effective – or remains inert and trivial when relevant related concepts are not subsequently developed during instruction. You can't judge a specific piece of biology knowledge such as a proposition that is being taught as trivial or fruitful, arbitrary or meaningful, mindless or mindful, unless you know the entire instructional design and what understandings will be constructed on the basis of it.

There are many kinds of concepts. The most "ordinary" are noun concepts. These are the easiest to learn, especially noun concepts that refer to concrete objects. Verbs are a special class of concepts that are used for linking two or more noun concepts together. Children and second language learners tend to master noun concepts before verbs. Concepts derive a composite meaning from the sum total of the connections among them that are created with linking words, usually verbs, that are sometimes called *relations* or *arcs.*

FAQ-3. WHAT CAN WE LEARN FROM THE HISTORY OF MAP MAKING (CARTOGRAPHY)?

An article called "Concept mapping and the cartography of cognition" (Wandersee, 1990) uses examples from the history of cartography to teach the nature of maps – both their strengths and their limitations. It develops the assertion that "to map is to know" – showing that we consider unmapped territory to be *terra incognita.* It also tries to show that cognitive mapping can be improved by considering what we know about geographic and thematic maps (maps with data layered on top of geographic features, such as weather maps, prevailing wind maps, ocean current maps, and magnetic declination maps). A companion paper which builds on these ideas, is entitled "The Historicality of Cognition: Implications for Science Education Research" (Wandersee, 1992b).

The largest paper map ever produced was a 135-by-66-foot map of the world designed by the National Geographic Society. It was made of 720 paper tiles and covered an entire university basketball court, yet it was a world one-millionth as big as the real world. Since the third century B.C., we have known how large (in circumference) the world really is. It is so large, in fact, that we can think about it only with the aid of maps (Wilford, 1998). So vast, too, is the human brain's knowledge base that special tools are needed to capture and represent even parts of it. Such tools include concept maps or semantic networks.

The map was a major invention of human thought. "To map is to construct a bounded graphic representation that corresponds to a perceived reality" (Wandersee, 1990). However, "a map is a product of compromises, omissions, and interpretations" (Wilford, 1998, p. 17). Thus, *the map is not the territory.* It is important to be aware that every map sacrifices some detail, distorts some spaces, and misinterprets some of the available data. Yet if knowing is tantamount to making a mental map of the concepts and relationships one has learned, and if people think with concepts, then the better one's cognitive map, the better one is able to learn. If it is true that "maps, like faces, are the signature of history" (Wilford, 1998), then our evolving concept maps represent our conceptual histories. In some cases, concept maps may show a learner what he/she doesn't know – similar to King Louis XIV's angry comment to his mapmaker after seeing that the new, more accurate map showed France to be much smaller than previously envisioned, "Your work has cost me a large part of my state!" (Wilford, 1998, p. 27).

FAQ-4. WHAT IS A CONSTRUCT?

A construct is a higher order concept that has no direct tangible referent in the real world. For example, the concept "pressure" is defined as force per unit volume – hence it rests on one's understanding of the concepts of force and volume. Pressure is an example of a construct, as is cancer. These two constructs might be described as states. Quite often, constructs describe properties of objects such as density, fecundity, carrying capacity, and biomass. Or they may describe processes such as cellular respiration, evolution, osmosis, photosynthesis, genetic drift, homeostasis, and learning.

While we don't graphically differentiate constructs from concepts on a concept map, typically they will be found in the upper levels of the map's hierarchy. It is important for students to realize that although constructs are not directly observable, they are far from uncommon in biology. One must understand the underlying biological concepts if one is to grasp the meaning of such constructs. Many attempts to increase public understanding of science fail because the explainers neglect to "unpack" the basic concepts underlying important scientific constructs they invoke. For example, it appears that the public's tenacious resistance to consumption of irradiated food arises because people believe such foods are radioactive. This represents a failure of public understanding of important physical and biological constructs, probably exacerbated by fear of radiation, distrust of corporations, and lack of clear explanation.

FAQ-5. WHAT ARE THE PARTS OF A CONCEPT MAP CALLED?

Novakian concept maps are usually hierarchical. The concept at the top of the hierarchy is called the *superordinate concept*. All other concepts are either called *subconcepts* or *subconstructs* (if they are constructs). Concept maps also contain *examples*, which are included as exemplars for particular concepts. For instance, the species *Cyperus papyrus* is an example of the wetland plant taxon commonly called the sedges. The linking lines on a concept map are labeled with *linking words* – chiefly verbs and their modifiers which link concept to concept to form *propositions*. These, in turn, allow concept maps to be read from top down through their individual branches. Another feature of concept maps is their *cross-links* (also labeled with linking words), which are important integrative propositions that bridge across branches of the concept map. New and insightful *superordinate concepts* and new and insightful *cross-links* are two signs of a thoughtful student that teachers should become sensitized to detect.

FAQ-6. HOW DO YOU CONSTRUCT A CONCEPT MAP?

Imagine you encounter the following botany article (excerpted below) while reading *The New York Times* (Freeman, 1998):

A Shade of Difference

> For a plant protein that was dismissed by skeptics in the 1950s, phytochrome has come a long way. Because of it, growers who want the best and biggest tomatoes, strawberries, and other crops need to be as fussy as an artist over the color of the plastic mulch they put in their fields.
>
> The protein directs plant growth and development in response to different kinds of light. And because of that, growers need to concern themselves with a color they cannot see, the far-red light that is just beyond the horizon of human vision, as well as the normal colors. Green leaves reflect far-red light, so plants can be fooled into thinking that they have lots of neighbors – and competitors – if they are bordered by mulch that reflects plenty of far-red wavelengths. In response, the deceived plants put more energy into their above-ground growth, and that means bigger tomatoes or strawberries that mature faster.
>
> They taste better, too, said Michael J. Kasperbauer, a plant physiologist with the United States Department of Agriculture in Florence, SC, who has been manipulating lights to manipulate plants almost since the existence of phytochrome was proved in 1959. A patented red plastic mulch developed by Dr. Kasperbauer and Dr. P. G. Hunt and manufactured by Sonoco is being sold to growers and gardeners.
>
> Strips of plastic mulch have long been known to help conserve moisture in some situations, but now scientists' attention is on mulch colors – whites, yellows, blues, and greens – to see how they can enhance things like flavor and insect control.
>
> Want a better root crop, like turnips? Pick a mulch color such as orange, that reflects much more red light and little far-red. The plants, not fretting about competitors, put their resources into growing bigger roots. (Freeman, 1998, p. B16)

How does one go about mapping the previous botany article excerpt? Here are the steps and details:

1. Extract or import 12–15 concepts pertinent to the article's central content. For example, you might extract these concepts: phytochrome, plant protein, far-red, light, colored plastic mulch, plant resources, above-ground growth, plant competitor effects, orange, root growth, tomatoes, strawberries, turnips, USDA, 1959.

2. The article is judged to be about phytochrome by the map maker, so that will become the superordinate concept. Rank-order the remaining extracted concepts by putting the broader, more general, ones near the top of your list, and moving the more specific ones towards the bottom of the list. Your reordered list might then look something like this: light, plant resources, plant protein, colored plastic mulch, above-ground growth, root growth, plant competitor effects, USDA, 1959, orange, far-red, tomatoes, strawberries, turnips.

3. Write these concept labels on individual Post-It™ notes. Move and arrange them on a table top or smooth vertical surface until you have created a linkable hierarchy that you think captures the central knowledge structure of the article.

4. Prepare a handwritten or print version of your concept map. Be sure to label all the linking lines with linking words and put arrowheads on the connecting lines to show directionality. Make sure that distinct levels of hierarchy are visible.

5. Edit or revise your concept map as needed. Admire!

See Figure 8.2 for a possible version of the concept map for the phytochrome article.

Role of Phytochrome

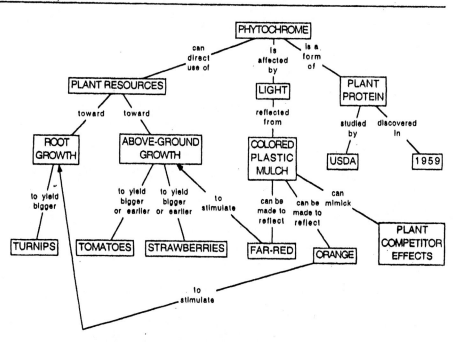

Figure 8.2. A concept map for the phytochrome article.

FAQ-7. WHAT GRAPHIC CONVENTIONS DO YOU RECOMMEND FOR CONSTRUCTING A CONCEPT MAP?

In harmony with Novak's original concept mapping practices, my research group has developed and for many years has used the following graphic conventions for drawing concept maps. This standard format for concept mapping was presented in Boston at the 1992 national convention of the National Science Teachers Association

(NSTA), by invitation of the President of the National Association for Research in Science Teaching or NARST (Wandersee, 1992a). Any map made to align with the "rules of common conversation" (a.k.a., standard graphic conventions) will be easily interpreted by anyone who is familiar with the system. This allows both teacher and students to focus on the *meaning* of the concept map, rather than having to initially decode its design. It is thus an aid to intelligent conversation about the ideas the concept map depicts.

The *Standard Concept Mapping Format* employs the following graphic conventions:

1. The completed concept map should have a single superordinate concept at the top of its hierarchy (accompanying modifiers are acceptable).
2. The completed concept map should resemble a branching tree root. Concepts should be placed at distinct and aligned levels of hierarchy within the mapping space that the map encompasses. Typically, concepts are arranged from general at the top to specific at the bottom, but the purposes of a particular map may call for exceptions to this arrangement.
3. A single concept map (which our research group calls a *micromap)* should be limited to about 12–15 noun elements (concepts, constructs, or examples). The purpose of this rule is to insure graphic effectiveness and to force prioritization of concepts. A *macromap* (a.k.a., map of maps) can be made to connect a set of *micromaps* at their upper levels of hierarchy. No concept map should be expected to capture all the knowledge in a domain, just the core. My research has shown that it can and does also serve as a visual stimulus for recall of many unmapped details. Again – the map is *not* the 'territory, it is a guide to it.
4. The text of each concept, construct, or example should be printed within its own box, circle, or ellipse.
5. Each linking line connecting two concepts should be labeled with linking words, so that the concept map can be read from the top down, through any of its branches. Two concepts with a labeled link between them form a unit called a *proposition.*
6. An example should be linked to another concept by the Latinate abbreviation, "e.g.," and should be enclosed by a dashed or broken box, circle, or ellipse. This practice is designed to make examples stand out from other concepts and constructs. Examples may be added at any level at the terminus of a branch.
7. Cross-links are used to represent important knowledge integration links across branches of the map. They are labled like ordinary links, but are drawn as dashed or broken lines, so that they stand out from the rest of the links.
8. Maps should be revised and redrawn so that there are *few or no crossed linking lines*. Graphic effectiveness demands that concept maps be as simple and direct as possible, with allowance for the limitations of the human brain's working memory capacity and the need for sufficient white space in such maps for the visual processing system.

An example of a concept map on rhizobotany is shown in Figure 8.3. Changes are needed to bring this map into alignment with the foregoing graphic conventions. Can you see what changes are needed?

FAQ-8. WHICH COMPUTER SOFTWARE CAN I USE TO CONSTRUCT A NOVAKIAN CONCEPT MAP?

The software application called *C-Map* was written for constructing Novakian concept maps via the Apple Macintosh computer (Stahl & Hunter, 1990). While that application does not facilitate the drawing of the dashed lines needed for representing cross-links and examples in the manner recommended in this chapter, one can easily modify the final print-out to conform, using correction fluid.

In addition to *C-Map*, concept maps of the type being recommended here can also be constructed using drawing, diagramming, or flow charting software packages available for both the IBM and Apple Macintosh platforms. *Inspiration* is a diagram construction package (developed by Inspiration Software, Inc., Portland, Oregon) that is quite popular for making concept maps and is available for both the IBM and Apple Macintosh platforms (Lanzing, 1998).

FAQ-9. HOW SHALL I GO ABOUT HELPING MY BIOLOGY STUDENTS BECOME PROFICIENT IN CONCEPT MAPPING?

As Adam Robinson (1993, p. 107) writes (metaphorically) in his book, *What Smart Students Know*, "the act of designing a...diagram will improve your understanding of information and help etch it on your brain." My own experience in teaching high school and college biology students how to concept map has shown that they first need to understand the following things about concept mapping.

1. Concept mapping was initially designed for use in the science classroom – although it can work well for studying other school subjects too. It is closely tied to a psychological theory of learning currently called *constructivism*.
2. Research has demonstrated that concept mapping helps students understand what meaningful learning is, and it appears to improve knowledge integration and retention. In addition, it can improve students' understanding of science, reduce students' anxiety levels, improve their perceptions about the subject matter, and can improve their performance on tests (Mintzes, Wandersee, & Novak, 1997, p. 428).
3. Students should be cautioned that it is normal to feel a little frustrated at first as they adopt a new way of learning. They should not be discouraged when mapping seems difficult initially, because it has long-term benefits. It takes at least eight weeks of using concept mapping before students become comfortable with it as a learning strategy and improve their performance in a course.

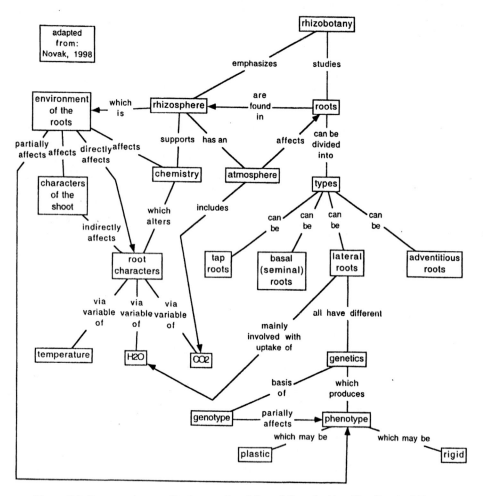

Figure 8.3. Concept map on rhizobotany that fails to follow the Novakian Standard Concept Mapping Format

4. Concept mapping is based on psychologist David Ausubel's assimilation theory of meaningful learning (Ausubel, Novak, & Hanesian, 1978). It has a

basis in what is known about how human memory works. About 200 research studies have been conducted regarding the effectiveness of concept mapping.

5. It is best to start by mapping a small topic that one knows well. It can be helpful to show students a concept map for an object they probably have in their hand at the moment, namely, a common pencil (see Wandersee, 1990, p. 933 for a map of this topic). It is also valuable to have students construct their first concept map with a partner, so they can talk over what goes where, and remind each other of the graphic conventions that we use for making a concept map (see FAQ-7.).

6. Students can draw their maps with erasable, colored markers on clear plastic overhead transparencies. In this way, students can share their maps by projecting them in the classroom. An attitude can be cultivated in which students are comfortable in sharing their knowledge with others and in negotiating the meaning of ideas with peers and the teacher.

7. Based on my experiences teaching science students and faculty from Brazil, the US, Canada, Germany, Great Britain, Finland, and Spain, my rule of thumb is that an individual needs to generate about 10 concept maps under the guidance of a veteran in concept mapping before he or she is ready to map scientific ideas efficiently, or ready to teach the strategy to someone else.

For many years, this author has taught a 3-credit-hour graduate seminar at Louisiana State University entitled *Concept Mapping.* This course devotes an entire semester to helping participants (primarily from the sciences, social sciences, and mathematics) construct and improve (in class) 10 different concept maps based on a variety of content sources, from biology videotapes to textbook sections to journal articles. The class also uses research papers, three textbooks on learning, and a large-scale concept-mapping project in its instructional design.

FAQ-10. HOW SHALL I EVALUATE MY STUDENTS' CONCEPT MAPS?

There are many proposed ways of evaluating concept maps to be found in the science education literature. Trowbridge and Wandersee (1994, 1996) have used a qualitative criterion-referenced checklist to give feedback on mapping performance to college biology students. It includes such aspects as checking: (a) the scientific validity of the map's propositions; (b) the labeling of linking lines; (c) hierarchical, dendritic structure; (d) presence of the 5 seed concepts supplied by the instructor; (e) presence of viable examples; (f) presence of important cross-links; (g) graphic effectiveness; and (h) conformity to mapping conventions. Some researchers have also used criterion maps made by specialists or experts as evaluative benchmarks.

Lavoie (1998) used the quantitative scoring rubric designed by Wallace and Mintzes (1990). This involves giving 1 point for each valid concept, 1 point for each valid connection between concepts, 1 point for each branch of the map's hierarchy, 1 point for each valid example or analogy, 5 points for each valid level of hierarchy, and 10 points for each important cross-link.

FAQ-11. CAN YOU DESCRIBE VARIOUS WAYS THAT CONCEPT MAPS CAN BE USED?

Concept maps can be used in the following ways and more. The concept map can

1. help the biology learner monitor his/her own learning,
2. help the teacher to see the effects of teaching on learning, and to negotiate concept meaning with the learner,
3. serve as a learning evaluation tool for formative or summative assessment, provided the learner is already competent in concept mapping,
4. be used as a research tool to investigate student cognition in science education,
5. be used as a curriculum planning tool,
6. be used to plan a lesson,
7. serve as a teaching tool to help students organize their knowledge,
8. be used to document conceptual change across instruction,
9. can be used to capture the knowledge presented in a research article, a television program, a textbook chapter, or a lecture, and
10. be used as knowledge interface on an Internet web site.

FAQ-12. WHAT DOES THE SCIENCE EDUCATION RESEARCH LITERATURE HAVE TO SAY ABOUT CONCEPT MAPPING?

While there are hundreds of concept mapping studies in the science education literature, the studies run the range from entirely qualitative to entirely quantitative. Experimental treatments vary immensely as do the operational definitions of what constitutes a concept map. This makes comparisons across studies quite difficult.

The only meta-analysis performed to date (Horton, McConney, Gallo, Woods, Senn, & Hamelin, 1993) found that concept mapping has moderate positive effects on student achievement and large positive effects on student attitudes.

Trowbridge and Wandersee (1994, 1996) have investigated the use of concept mapping in university biology courses. Their articles suggest how biology instructors can use this approach, even in large classes, and that instructor feedback gained from evaluating students' concept maps has the potential to dramatically improve the quality of instruction in a biology lecture course.

Perhaps the biggest flaw in concept mapping research is the failure to dedicate sufficient time to assure that the students actually become proficient in concept mapping prior to collecting research data. As noted above, proficiency requires practice and constructive feedback across the making of at least 10 maps and approximately two months of school.

It has yet to be determined what is the ideal sampling interval, what sample size is needed to assess students' knowledge development, and how stable a student's maps are over time, even when no instruction intervenes.

FAQ-13. WHAT ARE SOME OF THE CONTRIBUTIONS TO CONCEPT
MAPPING DEVELOPED BY YOUR 15° LABORATORY GROUP?

Information about the mission and work of the 15 Degree Laboratory, currently housed at Louisiana State University, may be found at the following URL: http://www.15degreelab.com. The Laboratory has developed such modest innovations as:

    (a) coconstructed concept maps (wherein a student's knowledge of a topic is concept-mapped in real time by a researcher conducting a clinical interview and the resulting map is amended and validated by the student being interviewed (Wandersee & Abrams, 1993),

    (b) iconic concept maps (Trowbridge & Wandersee, 1996) wherein scientific icons are used to depict complex objects or events, and concept mapping template paper (Figure 8.4), a lightly printed paper template consisting of levels of ellipses that can be darkened in, connected, and used selectively by students to produce a neatly arrayed and customized concept map (developed by Rosa Leathers, Kay Butler, and Jim Wandersee).

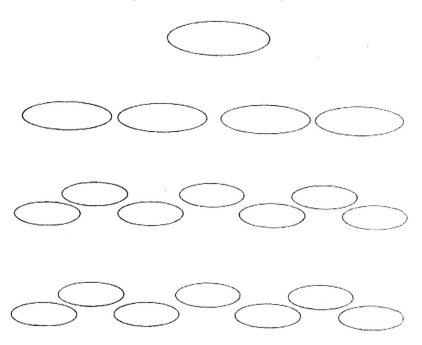

Figure 8.4 shows a piece of unused concept mapping template.

Figure 8.5 shows a student's concept map (rendered on a concept mapping template) for an article about the biomechanical challenges that had to be overcome when engineering a new humanoid robot (Thomson, 1998).

Concept mapping template paper, a relatively simple thinking aid, seems to make concept mapping an order of magnitude easier for many people. We speculate that this may be due to the power of selection from a limited set of visual options versus the "paralysis of analysis" that often results from the endless possibilities of the blank page or blank video mapping screen.

FAQ-14. WHERE CAN I FIND A BIBLIOGRAPHY ON CONCEPT MAPPING RESEARCH?

The reader is encouraged to look at research reviews of concept mapping (e.g., Horton et al., 1993; Novak & Wandersee, 1990; Ruiz-Primo & Shavelson, 1996), or at reference article lists (Al-Kunifed & Wandersee, 1990), or at selected research articles (Trowbridge & Wandersee, 1994, 1996; Wallace & Mintzes, 1990). The bibliography found in Novak (1998) is also helpful. Be forewarned that many graphic devices have been dubbed "concept maps" – therefore it is important to check to see if the article actually deals with Novakian concept maps, which have received the most research attention. Even if the study claims to use Novak-style maps, the maps may not be hierarchical or may omit using linking words – a fatal flaw, in my judgment.

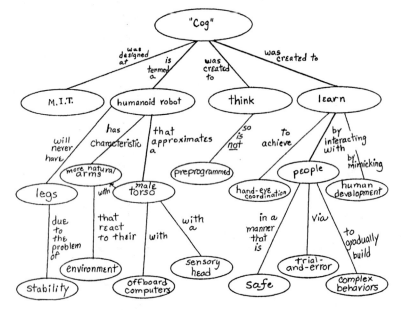

Figure 8.5. Student's concept map of an article about engineering a humanoid robot (Thomson, 1998).

FAQ-15. WHEN SHOULD CONCEPT MAPPING BE USED, GIVEN THE OTHER GRAPHIC METALEARNING TOOLS AVAILABLE?

An article by Wandersee (1990) and a book chapter by Trowbridge and Wandersee (1998) have presented the biology instructor with a number of options when choosing metacognitive tools. In general, concept maps are good at capturing the structure of a particular piece of knowledge, externalizing this knowledge, and then sharing it with others – even facilitating the creation of new knowledge. It is not the tool of choice for depicting processes, displaying cycles, guiding investigations, and so forth. We must assess both the quantity and the quality of the learner's knowledge if we are to guide the learner's education in an optimal way. We should also check for cognitive flexibility (multiple perspectives) and for connections across domains of knowledge – we dub the ability to do this kind of fluid knowledge integration *wisdom*, and it is still quite rare.

SUMMARY

Metacognitive tools such as concept maps are central to the success of a constructivist approach to learning science. Concept mapping is a form of graphic representation invented by Novak and his students, who saw it as a vital metalearning strategy. "To map is to know" – and unmapped territory may seem quite threatening to students – terra incognita. There are multiple ways in which concept maps can be used in biology classrooms.

Many kinds of concepts are used in biology. Object concepts are the easiest to learn and are typically nouns that refer to concrete objects (e.g., bone). Event concepts center on natural phenomena and involve object concepts arrayed spatially and temporally (e.g., meiosis). Property concepts are usually adjectives that serve to modify object or event concepts (e.g., dominant). In concept mapping, verbs with prepositions are used for linking two or more object or event concepts together (e.g., synthesized by). A construct is a special, higher order concept that has no direct tangible referent in the real world (e.g., gradient). Constructs are usually the most challenging ideas to teach and learn.

Novakian concept maps are hierarchical by design. The concept at the top of the hierarchy is called the superordinate concept. All of the other concepts are called subconcepts and are connected by labeled linking lines. Concept maps may also contain anchoring examples and cross-links that represent knowledge integration. Graphic conventions in the Standard Concept Mapping Format have been described. A rule of thumb is that one needs to construct about 10 concept maps (and have them critiqued by a knowledgeable biologist experienced in concept mapping) before one can be considered a proficient biology "mapper" who is ready to teach the strategy to someone else. The strategy typically takes students at least eight weeks of a school term to master.

Evaluation of a concept map involves checking the: (a) scientific validity of the map's propositions; (b) labeling precision of linking words; (c) hierarchical, dendritic map structure; (d) presence of assigned seed concepts; (e) presence of viable examples (preferably novel ones); (f) presence of biologically significant cross-links; (g) graphic effectiveness of map's layout; and (h) conformity to Standard Concept Mapping conventions.

The only meta-analysis performed to date found that concept mapping has moderate positive effects on student achievement and large positive effects on student attitudes. Long-term studies are needed using data drawn from students with demonstrated proficiency in concept mapping using Standard Mapping Conventions.

KATHLEEN M. FISHER

CHAPTER 9

SemNet® Semantic Networking

Having Fun!

SemNet® is both user friendly and fun to use. The tutorial took the better part of a day (5 or 6 hours) but we suspect this was because we were having fun and got carried away. The tutorial is quite comprehensive, giving you a very good introduction to the program. The manual is a marvel for a "small unfunded team of five people all holding full-time jobs doing research and teaching, much of it not directly related to SemNet®." . . . Though neither slick nor professional looking, it is clear, complete, and easy to find stuff in (with a great index) and has lots of helpful suggestions (Weitzman & Miles, 1995, p. 308)

SEMANTIC NETWORK THEORY

"When all fields of knowledge are considered, semantic networks are probably the single most pervasive form of knowledge representation used today – a semantic network underlies every sentence in natural language and every coherent logical description" (Lehman, 1992). Almost every application area in artificial intelligence (AI) has used semantic networks, from machine vision to natural language understanding to information retrieval to dynamic control of combat aircraft (Lehman, 1992; Lehman & Rodin, 1992).

Quillian (1967, 1968, 1969) showed us how to capture human semantic structure and processing in a computer. As Collins and Loftus (1975) describe it:

Some years ago, Quillian . . . proposed a spreading activation theory of human semantic processing that he tried to implement in computer simulations of memory search . . . and comprehension . . . The theory viewed memory search as activation spreading from two or more concept nodes in a semantic network until an intersection was found. The effects of preparation (or priming) in semantic memory were also explained in terms of spreading activation from the node of the primed concept. Rather than a theory to explain data, it was a theory designed to show how to build human semantic structure and processing into a computer.

There is an enormous body of research on semantic networks in cognitive science, psychology and artificial intelligence (e.g, Sowa, ed., 1983, 1990, 1999; Brachman & Levesque, 1985; Brachman, Levesque, & Reiter, eds., 1991; Jonassen, Beissner & Yacci, 1993), as well as on spreading activation theory (e.g., Collins & Loftus, 1975). The validity and utility of the semantic network model have been well accepted

143

across a number of fields of study for more than a quarter of a century, and semantic network theory is supported by extensive research and practice. There is a small but growing literature on semantic networks in education and this is a field rich in research opportunities. Semantic network theory is consistent with the learning models proposed by learning theorists Ausubel (1963, 1968), Novak and Gowin (1984) and especially when used to support group learning, Vygotsky (1978).

The review below briefly describes research in various fields to illustrate the richness of available information. At the same time, the chapter space severely limits what can be said about each individual research study.

SEMNET® KNOWLEDGE MAPPING

Knowledge mapping is the representation of detailed, interconnected, nonlinear thought (Fisher & Kibby, 1996). A knowledge map is an external mirror of your own radiant thinking that gives you to access your vast thinking powerhouse (Buzon & Buzon, 1993). Knowledge mapping is an external extension of working memory which especially supports reflective thinking (Perkins, 1993; McAleese, Grabinger and Fisher, 1999). Knowledge mapping promotes comprehension skills well beyond simple decoding.

SemNet® is a knowledge mapping tool that draws upon a combination of semantic networking theory and computer technology to change the ways in which students engage in science learning, shifting the emphasis from rote memorization to meaningful understanding. Learners can use the SemNet® software to communicate aspects of their thinking to their teachers and vice versa:
"Oh, I see what my student is thinking!"
"Oh, I see what my teacher is thinking!"
Thousands of semantic networks have been constructed about many knowledge domains by individuals of many ages. For example, I taught SemNet® to my four-year-old granddaughter one day, and later found her teaching it to her three-year-old friend. As indicated in the opening vignette, beginners can construct knowledge nets after a brief introduction. At the same time, the fun and fascination persist even after years of use. Long-time power-users of SemNet® enjoy the fact that we can still teach one another new tricks.

SemNet® has been used as
- a *learning tool* that supports personal and group knowledge construction,
- a *knowledge analysis tool* that helps users unpack complex ideas such as a chapter in a text,
- a *knowledge presentation tool* both in the classroom and on the web,
- a *qualitative research tool* for eliciting and mapping individuals' ideas,
- a *knowledge resource* in which learners can look up information,
- a *tool for mapping feelings*, and
- a *succinct communication tool* for working with people from different language backgrounds.

A hidden strength of SemNet® is that when teachers review semantic networks constructed by students, they are able to give *specific, targeted diagnostic feedback* to facilitate students' learning. For example, second-language biology learners have considerable difficulty with the verb relations that tie concepts together. Many of my students have trouble discriminating between the two opposing rays of the basic whole/part relation, *"has a part "* and *"is a part of."* As a consequence, they are enormously disadvantaged in learning biology, since biology teachers and students use this particular relation in more than 20% of all the propositions they generate. Correcting a failure to understand a relation is much more powerful than correcting any single concept, since relations are used over and over again.

We are not the first to discover that second-language learners learn nouns before verbs (Rosenthal, 1996). The same pattern is seen in children learning their first language (Gentner, 1978, 1981a, 1981b, 1982). The good news is that when second-language biology students construct semantic networks of their biology knowledge as they are learning, and they are given diagnostic feedback, their mastery of verb phrases increases rapidly. This level of diagnosis and feedback is not easily achieved without a tool such as SemNet®.

MIRROR OF THE MIND

Thro (1976, 1978) studied the relationships between the cognitive structures of learners in a college physics class and the physics being taught. She observed that, over time, students' mental models became more and more like the instructor's mental model. She also noted that students' problem-solving ability was directly related to their differentiation among the clustered items in their cognitive structures. This close relation between evolving cognitive structure and performance has been observed in other studies as well (Schvaneveldt, 1990b; West & Pines, 1985).

Gordon and Gill (1989) found that, by mapping student knowledge in two domains, they were able to predict student performance with 85% and 93% accuracy. They used conceptual graphing (a variant of a semantic network) to analyze students' declarative knowledge. The two domains they examined were mathematical vectors and how to use a videotape recorder. Their observations, that skilled performance derives from detailed, well-organized, declarative, domain-specific knowledge, are now widely accepted. Their knowledge elicitation methods are described by Gordon (1989, 1996). In a similar vein, Davis, Shrobe & Szolovits (1993) claim that "Representation and reasoning are inextricably intertwined" (p. 29).

RICH ENVIRONMENT FOR ACTIVE LEARNING

Rich Environments for Active Learning or *REALs* (Grabinger, Dunlap, & Duffield, 1997; Grabinger, 1996; Grabinger & Dunlap, 1995) are computer-based environments that give students increased control over the learning process while they acquire both content knowledge and life-long learning skills. Similarly, a knowledge arena is a virtual space in which learners can operate on ideas (McAleese,

1985). The SemNet® software can provide both a Rich Environment for Active Learning and a knowledge arena for operating on ideas.

CONNECTIVITY

"The task of learning is primarily one of relating what one has encountered (regardless of its source) to one's current ideas" (Strike & Posner, 1985, p. 212). A semantic network consists of arrays of old and new ideas (concepts) linked by named relations (Figure 9.1). The connectedness of semantic networks is the key to their power in learning and retrieval. For example, in a list of 2463 concepts, the distance from the first to the last node is *2462 nodes*. In contrast, in our semantic network in biology containing 2463 concepts (Figure 9.2), the longest "shortest path" between two concepts is just *11 nodes!*

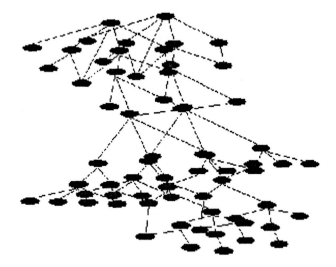

Figure 9.1. Portion of a semantic network.

Interconnectivity provides enormous power for information processing and retrieval in both human memory and in computer-based semantic networks. See Horn (1989) for further discussion of semantic network-style hypertext, interconnectivity, and their applications in the business world.

Structured connectivity is achieved in a semantic network by building relations between concepts in a systematic and precise way. Precision of meaning requires using named relations as links. There is an ongoing tension between parsimony and abundance of relations used to describe a thing or event. Parsimony promotes ease of network construction and retrieval of ideas. On the other hand, a principled abundance of relations can allow the net-builder to make finer and finer discriminations between ideas, while an undisciplined abundance of relations can simply contribute to confusion within a knowledge representation.

Networks are rich and irregular structures that often contain embedded hierarchies (e.g., parts of a flower), temporal flows (e.g., sequence of events in mitosis), causal chains (e.g., production of spindle fibers by centrioles), and other knowledge substructures. Many of these knowledge structures can be "pulled out" of a network for examination as described below.

THE KNOWLEDGE CORE

Another way of demonstrating interconnectivity in SemNet® is to view the knowledge core that consists of the 49 most *embedded* concepts in a network arrayed in a spiral (Figure 9.2). If an entire network is compared to a mountain chain, then when we look at the most embedded concepts we are looking at the mountain peaks – the most important, unifying interconnected ideas. Figure 9.2a shows the *direct connections* from the most embedded concept (explained below) to the other 48 highly embedded concepts in a knowledge network about biology containing a total of 2463 concepts. Two-step paths in the knowledge core from the most embedded concept to *two nodes away* are shown in Figure 9.2b. Langer (1989, 1997) emphasizes that multiple perspectives within a knowledge structure are valuable in learning and mastery.

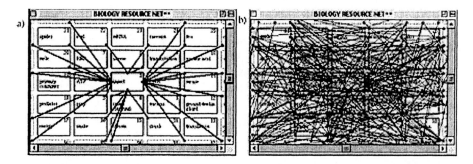

Figure 9.2. The inner part (first 25 concepts) of the 49-concept knowledge core in the Biology Resource Net. a) The direct links from <u>insect</u>, the most embedded concept, to the other highly embedded concepts; b) the unique paths from <u>insect</u> to two nodes away. This net contains 2463 concepts, 83 relations, and 4187 instances.

EMBEDDEDNESS OR MAIN IDEAS

Embeddedness is determined by the number of unique paths from a concept to two nodes away, as illustrated in Figure 9.3. <u>Concept I</u> is directly linked to (participates in instances with) four other concepts called <u>A</u>, <u>B</u>, <u>C</u>, and <u>D</u>. Each of those four concepts is linked to three other concepts (numbered 1 through 12). Each of the twelve numbered concept is two nodes away from Concept I. Therefore, there are twelve unique paths from <u>Concept I</u> to two nodes away.

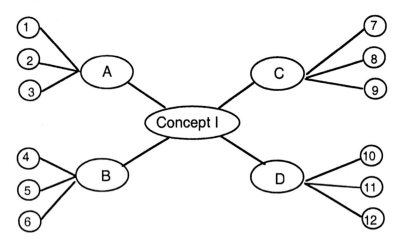

Figure 9.3. Concept I with an embeddedness of 12. This map contains 17 concepts, 1 unnamed relation, and 16 instances.

The valuable thing about embeddedness is that it automatically identifies the main ideas in any network describing a knowledge domain. In our net corresponding to an introductory biology course, the twenty-five most embedded concepts look like the course syllabus. They are the main ideas in the net, and they appear at the top of the display of Concepts by Embeddedness.

CASCADING DOWN A HIERARCHY

Research shows that some forms of knowledge organization are more effective than others in facilitating recall, application and problem solving. Systematic hierarchical organization is especially useful in the sciences (Reisbeck, 1975; Reiger, 1976; Reif & Heller, 1981, 1982; Reif & Larkin, 1991). SemNet® reinforces such organized thinking in part by allowing the user to construct and extract hierarchies, causal chains, temporal flows and other internal structures from a web for review and reflection.

For example, in a food web net describing a forest community, the hierarchy of what a bear eats (both directly and indirectly) is five pages long. An excerpt of this hierarchy is shown in Figure 9.4. The food web hierarchy captures the flow of matter and energy through the forest community. Additional relations are needed to capture other relationships among the interacting organisms.

bear
   eats mouse
          eats   nut
          eats   seed
          eats   herb
          eats   fruit
          eats   millipede
                    eats   grass
                    eats   herb
                    eats   decaying matter
          eats   insect
          eats   spider
                    eats   millipede
                    eats   insect
                    eats   centipede
                              eats   insect
                              eats   insect (preying)
                                          eats   millipede
                                          eats   insect

Figure 9.4. A small portion of the hierarchy extracted from a food web net. A hierarchy is automatically extracted from a net by SemNet® when the user requests that the hierarchy begins with a particular concept and follows one or more relations. This net contains 39 concepts (types of organisms), one relation (eats/eaten by), and 233 instances (concept-relation-concept units or propositions such as "centipede eats millipede").

Other knowledge structures may involve temporal flows such as the flow of blood through the body, as in a simplified net about the circulatory system (Figure 9.5). Constructing even such simple-minded flows as the one below seems to help students acquire a deeper understanding of a dynamic process.

Figure 9.5 also shows how to create tripartite relations in SemNet® (that is, relations among three concepts), even though the software is designed to represent only bipartite relations (relations between two concepts). The essential trick is to incorporate one of the concepts (in this case, <u>blood</u>) into the connecting relation, *passes blood to.*

```
                                    receives blood from          heart
                          receives blood from          artery
                receives blood from          capillary
           receives blood from          vein
 heart      °
           passes blood to    artery
                passes blood to    capillary
                     passes blood to    vein
                          passes blood to    heart
```

Figure 9.5. Following the bidirectional relation, receives blood from/passes blood to, from the concept, heart, in a net called "Heart 2." This net contains 138 concepts, 25 relations, and 205 instances.

The most frequently used hierarchies in introductory biology involve categories of things, as in "an <u>animal cell</u> *has organelle* <u>nucleus</u> which *contains* <u>chromosomes</u> which are *composed of* <u>DNA</u>" (Figure 9.6). This is an example in which finer discrimination is perhaps achieved by using three separate relations *(has organelle, contains,* and *composed of)*, whereas if parsimony were to be invoked, the relation *"has part"* could be used for all instances shown in Figure 96.

```
 animal cell
           has organelle       cytoplasm
           has organelle       Golgi Body
           has organelle       lysosome
                     contains enzymes
           has organelle       mitochondria
           has organelle       nucleus
                     contains chromosome
                     composed of       DNA
                     contains genes
                     contains hereditary information
```

Figure 9.6. A hierarchy beginning with the concept, animal cell, and following three flagged relation rays, has organelle, contains, and composed of. From the Cell Exp net containing 163 concepts, 19 relations, and 260 instances.

Complex hierarchies involving multiple relations can be automatically extracted from a semantic network. Such a hierarchy or tree network possesses two important and fundamental properties – inheritance and recognition (Garvie, 1994). Inheritance reduces the learning load since it is possible to learn the features shared by all members of a category (such as an <u>organelle</u>) just once, and then apply these features to all members of the category. Use of hierarchies also helps learners to structure their knowledge for easy retrieval.

CATEGORIES AND FUZZY SETS

Hierarchies are built with categories. The importance of categories as thinking tools cannot be underestimated (Rosch, 1973, 1975, 1977; Rosch & Mervis, 1975; Rosch, Mervis, Gray, Johnson, & Boyes-Braem, 1976; Rosch & Lloyd, 1978; Lakoff, 1987). They are intrinsic elements in our most basic thought patterns, are generated spontaneously, automatically and subconsciously, and have much more interesting and complex organization than was first imagined. For many decades, it was believed that categories had distinct edges and contained elements having a variety of properties in common. Now we know that categories have fuzzy edges more often than not and contain members ranging from prototypical to fringe. Furthermore, there is often no common set of properties shared by all members of a category (Rosch, 1973, 1975, 1977; Rosch & Mervis, 1975; Rosch, Mervis, Gray, Johnson, & Boyes-Braem, 1976; Rosch & Lloyd, 1978; Lakoff, 1987). Lofti Zadeh's fuzzy set theory (Zadeh, 1963, 1973, 1976, 1979) provides mathematical ways of describing categories with fuzzy edges. Fuzzy logic has a variety of practical applications (McNeill & Freiberger, 1993).

Using categories effectively is a key to good biology learning. If we have learned about organelles, and if we are told that a mitochondrion is an organelle, then we know that a mitochondrion is a subcellular structure that performs a specialized task and is surrounded by a membrane (although some biologists subscribe to a broader definition of organelle which includes structures with and without membranes). Organizing related ideas into categories saves a lot of repetitive learning.

Each member of a category theoretically inherits the distinguishing properties of that category. However, since category structure has been found to be much more complex than originally thought, it is necessary to learn the exceptions or the fringe members of the category, such as birds that don't fly.

Nearly everything we talk about in biology is a category. Macromolecule, DNA, organelle, mitochondrion, dog, Collie, and population are all generic ideas or categories. SemNet® construction provides a means for helping students use categories systematically and effectively in their thinking. Differentiation is the key. Many students do not discriminate very well at first between closely related concepts such as oxygen atom, oxygen molecule, and oxygen gas (all called "oxygen" for short), or between eukaryotic cell, animal cell, and plant cell. Constructing semantic networks definitely facilitates making these finer distinctions.

CHUNKS OF INFORMATION

As you can imagine, it is a challenge to represent a complex interconnected knowledge structure in a comprehensible manner. The SemNet® solution is to look at one concept at a time in its graphic frame, with all its links to other concepts (Figure 9.7). This has the advantage of providing not only a coherent graphic frame about that concept, but it provides a manageable chunk of information that is believed to be more easily assimilated than a text-based description.

Chunks were first described by Chase and Simon (1973) in studying chess players. An initial assumption was made that master chess players have tremendous calculation abilities and can look at a chessboard and figure out what will happen five moves ahead. What Chase and Simon discovered instead was that master chess players have well-developed pattern recognition abilities. Chess players quickly recognize the arrangement of pieces on a chessboard and associate certain outcomes with each arrangement. The pieces on the chessboard comprise a chunk of information that is perceived as an image and converted into a schema. In general, a schema is a strategy by which the mind organizes elementary units into larger patterns.

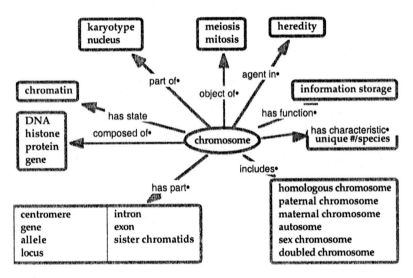

Figure 9.7. A single frame from the Biology Resource Net. The central concept in this frame, <u>chromosome</u>, is linked to 25 related concepts by eight different relation rays.

Simon (1974) estimated that a class A player has a repertoire of about 1,000 such schemata, while a master chess player has between 25,000 and 100,000 chess board schemata stored in memory – about the same range as an educated person's vocabulary. The SemNet® graphic frame provides a chunk of information about a concept.

DUAL CODING

According to dual-coding theory (Mayer & Sims, 1994), the learner can encode information in two distinct information processing systems, one that represents information verbally and one that represents information visually. Dual coding facilitates ability to both retrieve and apply ideas (Mayer & Sims, 1994). As Charles S. Peirce (1976, 1931–1958), the father of semiotics, has said, all thought is diagrammatic. There is substantial evidence to suggest that images generated by the spatial processing system are malleable but relatively long-lasting (Shepard & Cooper, 1982).

There is a graphic frame for every concept in a semantic network. Each frame provides a visual image, and each can be read as a series of verbal propositions. Thus, a semantic network promotes dual coding of information which in turn promotes retrieval. The graphic image also provides a concrete arena for comparing, contrasting and manipulating verbal propositions (Amlund, Gaffney, & Kulhavy, 1985).

SemNet® graphic frames seems can also include a picture or other informative image as in Figure 9.8. Liu (1993) suggests that such images can be important learning aids for low achieving students, since low achieving students in her study used graphics significantly more frequently than the two higher achievement groups.

We have also observed that the images can help nonbiologists understand biological ideas. For example, a cognitive scientist who was struggling to understand the genetic term, translation, had an "aha" experience upon seeing the image of translation in the figure below. Kozhevnikov, Hegarty, and Mayer (1999) now have evidence that the mind codes information in at least three ways, verbally, spatially, and in pictures.

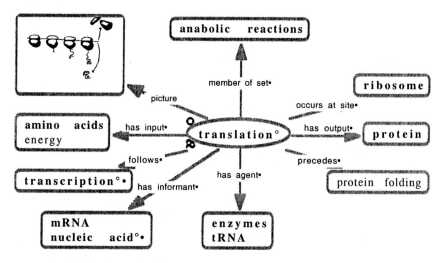

Figure 9.8. A graphic frame containing the central concept, translation, with an attached picture and 11 related concepts. From a Bio Starter net with 167 concepts, 24 relations, and 276 instances.

BUILDING CORE BIOLOGY KNOWLEDGE

More than half (>50%) of all the instances entered into semantic networks reflecting introductory biology knowledge use just three relations: *set/subset, whole/part,* and *characteristic.* This was first observed in an analysis of 3921 instances created by nine students enrolled in an introductory biology course for majors (Fisher, 1988). The same usage pattern has been observed many times since. This observation holds true whether representations are created by students or faculty. These three relations appear to account for our basic understanding of introductory biology concepts.

Despite the widely shared meanings, however, there is not a widely shared nomenclature. A *whole/part* relation can have many different names such as *has a part/is a part of, is composed of/is a component in, contains/ contained in,* and so on. That is, relation use is consistent, but relation names are idiosyncratic.

In contrast, in a domain such as biology, an entire language of concepts develops with a high level of agreement among biologists regarding the meaning of each concept. The set of shared relation names is much smaller. Our talking and writing are made more interesting and our thinking more fluid, perhaps, by *not* standardizing relation names the way we do concept names.

Some people use the *whole/part* relation to refer to physical structures and their parts (as in tissue *is composed of* cells). Others use it to refer to a process and actors in that process (as in glucose *is part of* glycolysis). Still others use it to refer to sets and their members (as in dog *is part of* mammals) (Fisher & Faletti, 1989). For the sake of clear thinking, I strongly encourage my students to be consistent in using the *whole/part* relation to describe *only physical wholes and their parts.* This is a more specific use than the definition proposed by Aristotle (Hope, Translator, 1997/1952).

DOMAIN SPECIFIC KNOWLEDGE

Although three relations are used 50% of the time, other relations play critical roles in capturing meaning. Even though most relations are used far less frequently than the three above, the particular meanings they capture can be essential for the accurate representation of the subject.

Luoma-Overstreet (1990) kept a journal as she used SemNet® to map the information in John Anderson's (1983) book, *The Architecture of Cognition.* One of her goals was to determine if there is such a thing as a perfect set of relations. She found that the "perfect set" of ten relations she began with only lasted a few pages into the first chapter, and that new and different relations were required for every chapter. She concludes: "It's obvious to me now that *content drives relations."* (italics added).

In another study, Hoffman (1991) conducted telephone interviews with experienced SemNet® users scattered around the world, to examine the ways in which they created and used relations in their nets. Supposing, he said, you were limited to using just seven relations in constructing your nets – which seven relations would you choose? The interviewees played along and chose their seven favorite relations, but felt that they would be *severely limited* by the exclusive use of their own personally chosen relation set.

He then compared the seven relations chosen by each interviewee with his preferred seven relations. There was strong agreement on the three "ubiquitous" relations, *set/subset, whole/part,* and *characteristic,* and modest agreement on the *cause/effect* relation. However, the three remaining relations were largely unique to each user. Since Hoffman's interviewees were working in many different fields, this again emphasizes that *content (and perhaps personal preference) drives relations.* It also suggests that the three ubiquitous relations play an important role in understanding well beyond the domain of biology.

Cooke and McDonald (1986) studied the Pathfinder software (Schvaneveldt, 1990b) as a tool for eliciting expert knowledge. Pathfinder constructs a network of ideas interconnected by unlabeled lines, based upon similarities or differences perceived by the subjects. Cooke and McDonald (1986) concluded that Pathfinder has limited utility for eliciting expert knowledge because of the absence of named relations. The nature of each link must be specifically identified in order to capture meaning.

The main conclusion, then, is that *specialized knowledge requires specialized relations.* A constrained set of relations such as that favored by Holley and Dansereau (1984) severely limits the meanings that can be expressed. The total absence of named relations limits meaning-making even further. The specific relations that are required are a function of the content and context. The names applied to those relations are determined in part by personal preference.

DISCRIMINATION AMONG IDEAS: BIG CITIES AND SMALL TOWNS

A knowledge network constructed with the SemNet® software is like a road map. A semantic network includes the largest "cities" (key concepts) as well as the smallest "towns" (subordinate, ancillary and anchoring concepts). It also shows many possible "roads" (links) among them. When concepts are listed by embeddedness, the "cities" (main ideas) are ordered by SemNet® from the largest to the smallest in terms of connectivity. "Cities" (concepts) can also be ordered from the largest to the smallest in terms of their direct links to other concepts.

Gorodetsky and Fisher (1996) found that when students use SemNet®, their ability to discriminate between main and subordinate ideas increases significantly (see also Gorodetsky, Fisher & Wyman, 1994). Students report that this enhanced ability to identify main ideas occurs not only in a course where SemNet® is being used, but transfers to other courses as well, resulting in altered note-taking methods. Similarly, Chmielewski & Dansereau (1998) observed that students who were previously trained in concept mapping recalled more macro-level ideas from text passages than students who had not received such training, even in situations where concept mapping was not used. These two studies provide evidence that mapping knowledge leads to enhanced cognitive skills that can then transfer to other domains.

MULTIPLE WAYS OF SEEING

SemNet® provides many different views of concepts, relations and instances. In fact, there are over 20 different ways of viewing the information in a single semantic network. These multiple views of a given knowledge structure facilitate learning (Langer, 1988, 1997). Further, as students develop facility in using different views to obtain different kinds of information about their knowledge structure, they gain cognitive flexibility (Spiro, Feltovich, Jacobson, & Coulson, 1991; Fisher, 1995; Fisher & Gomes, 1996a, 1996b), deeper content learning (Langer, 1988, 1997; Marra, 1996), and enhanced metacognition (Tobias & Everson, In Press; Fisher, 1993; Gorodetsky and Fisher, 1996; Fisher & Faletti, 1993).

HIGHWAYS AND BYWAYS: COGNITIVE FLEXIBILITY

Advanced knowledge acquisition refers to learning a content area beyond the introductory stage but before extensive experience and practice (Spiro, Feltovich, Jacobson, & Coulson, 1991; Spiro & Nix, 1990; Jacobson & Spiro, 1995). It differs in important ways from introductory learning. At the advanced stage of learning, knowledge must be active rather than inert and reasonably correct. The goals of learning shift from *knowledge reproduction* to *knowledge application*.

Cognitive flexibility theory was initially developed to describe the needs of advanced medical students (Spiro, Feltovich, Jacobson, & Coulson, 1991); Spiro & Nix, 1990; Jacobson & Spiro, 1995). It applies as well to students who are preparing to become teachers, as they near the end of their student careers (Fisher & Gomes, 1996a, 1996b). Even young students benefit when they acquire more active knowledge and less inert knowledge (Langer, 1998, 1997). SemNet® can promote cognitive flexibility by the scaffolding it affords for student knowledge analysis, construction, reflection, and revision; by the multiple representations it offers of each knowledge structure; by the complexity and irregularity of knowledge structures it allows; and by the concrete visualizations (knowledge arenas) it offers for operating on ideas.

MATRIX OF MEANING

A semantic network is a medium of expression for conveying skeletal ideas, the core meanings contained in a given unit of knowledge. A semantic network is also a tool for intelligent reasoning (Davis et al., 1993). It can capture and express both *knowledge about* (declarative knowledge), and to some extent, *knowledge of how to* (procedural knowledge). Although semantic networks as we depict them are word-based, the skeletal structure of the concepts (schemata) gets close to raw thought, as described by Pagels (1988, p. 23):

> It seemed obvious to me that if you want to understand a spoken language, you ought to study the people who speak it and speak it well. The simultaneous translators often employed by various state departments are masters of spoken language. One of these, a Soviet citizen, is truly remarkable in that he knows dozens of languages, Oriental as well

as Western. If you want to understand how languages work, this is the sort of person you ought to meet and study. After he listens to someone speak, he translates the remarks into whatever language is desired — any one of dozens. How does he do it? According to him, he "hears" the remarks not in any language at all, but rather as "a matrix of meanings" — a conceptual format of some kind that he creates. When asked to translate into a specific language, he consults the matrix and expresses that meaning into a language. It would appear that spoken language is subordinate to a nonverbal format, a deeper logical structure that is independent of any specific language. This seemed rather clear to only a few people in the early 1960s, although it has become better accepted today, most notably through the work of Noam Chomsky.

Pinker (1994), a linguist, likewise refers to the skeletal ideas with which we think and which are free from language. He calls them "mentalese." I propose that a SemNet® graphic frame captures some of this matrix of meaning or mentalese (Figure 9.9). The matrix of meaning involves images, associations, and nearest neighbors. Imagine you can zoom in on the rabbit in Figure 9.9 to look at his ears or his tail, or spin him around to see his back, or send him running — all in your mind, of course.

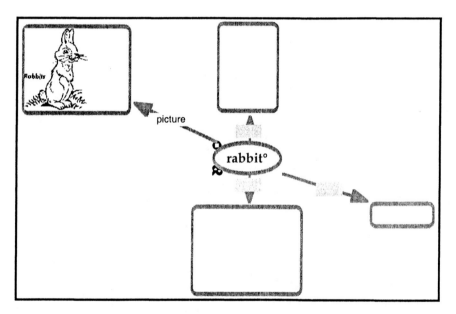

Figure 9.9. A matrix of meaning about the concept, rabbit.

The matrix of meaning in figure 9.9 could be filled in with words from any language (Figure 9.10). This interlingua concept was proposed by Richens in 1956 (cited in Lehman, 1992).

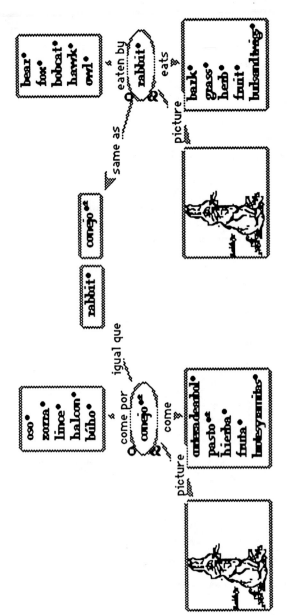

Figure 9.10. English/Spanish representation of the central concept, <u>conejo</u> (rabbit), in a food web net.

THE POWER OF NAMING

Words serve as handles for ideas that are captured in matrices of meanings. This link between a word and its skeletal meanings is what makes words representational (Brachman & Levesque, 1985). There is a correspondence between a graphic frame created in a semantic network, the matrix of meaning in the mind, and the thing in the external world. Macnamara (1984), who has written a book about the importance of names for things, begins with Genesis, which says (in one version) that man's very naming of the animals marks his dominion over them. With a name we have the beginning of *control over an idea, a place to file it, and a handle with which to retrieve it.* In a guided discovery classroom, students typically explore the ideas or phenomena first and then later attach names to them. SemNet® is used in my class to support students in making this shift from experiential to conceptual learning.

LOOKING THROUGH THE LANGUAGE BARRIER!

A SemNet®-like structure is uniquely suited to support cross-language learning and communication because of the power of this meaning matrix. With the English/Spanish semantic networks on the web (Fisher, 1999), it is possible to open two browser windows and click through the Spanish part of the network in one window while clicking through the corresponding English concepts in the other window.

Mapping has been studied as a promising tool to support language learning (Lambiotte, Dansereau, Cross, & Reynolds, 1989; Amer, 1994), even computer languages (Fegahli, 1991). Amer (1994) found, for example, that concept-mapping Egyptian students studying science in English wrote significantly better summaries than an experimental group that underlined text and than a control group using no special learning strategy. Hofmann & Welschselgartner (1990a & 1990b) in Germany used SemNet® semantic networks, primarily in English, to facilitate communication among the members of their multilingual European Common Market committee. Liu (1993) found that the scores of 63 English as a Second Language (ESL) students using semantic-network-based *Hypermedia-Assisted-Vocabulary-Learning Courseware* increased significantly from pre- to post-testing. Use of the system also resulted in improved ability to use the words appropriately in context. Thus a number of studies suggest that second language learning can be enhanced by knowledge mapping.

PROMOTING THINKING: COGNITIVE SKILLS

Gorodetsky & Fisher (1996) compared learning in two sections of a capstone course in biology for prospective elementary school teachers. Students in one section used SemNet® as a study tool, while those in the other studied by traditional means. Both groups responded to short essay questions at the end of the course. SemNet® students included significantly more (twice as many) relevant biology words in their responses, they wrote twice as many sentences, and their sentences were shorter and had greater clarity than those written by the traditional group.

Reader & Hammond (1998) in the UK compared one group of students using a Hypercard-based knowledge mapping system with a second group using a computer-based note-taking strategy. Both groups were learning the same material. At the end of the study, students took an achievement test that asked for both factual and relational information. The knowledge mapping group earned significantly higher post-test scores than the note-taking group. The authors concluded that computer-based mapping aided the acquisition of both relational and factual knowledge.

Christianson & Fisher (1999) examined the learning of diffusion and osmosis in a student-centered, inquiry-based nonmajor biology course in which students used SemNet® as a learning tool (the course is described in Fisher, in press b). Understanding of diffusion and osmosis by students in this course was compared to that achieved by students in two teacher-centered, lecture-based nonmajor biology courses in which SemNet® was not being employed. The assessment used a two-tiered test for conceptual understanding of osmosis and diffusion developed by Odom & Barrow (1995). Students in the inquiry-based SemNet® course achieved significantly higher scores, with the differences occurring primarily in comprehension of the underlying mechanisms of these two processes. This is consistent with findings from other studies that when students engage in carefully structured hands-on learning experiences combined with systematic knowledge representation, their content learning and meaningful understanding increase.

PROMOTING THINKING ABOUT THINKING: META-KNOWING

Metacognition has at least two distinct aspects. First, it refers to one's knowledge about one's own cognitive processes and products (Flavell, 1977). Construction of a semantic network necessarily increases one's awareness of one's own cognitive processes and products, since it entails making one's thoughts explicit and capturing them for reflection, discussion and revision.

Second, metacognition refers to the active monitoring and regulation of cognitive processes (Flavell, 1977). The regulation and orchestration of cognitive processes are a natural fallout of semantic networking. Revising and polishing a SemNet® network not only results in improving a particular knowledge map, but also alters the ways in which the net-builder thinks about thinking and engages in thinking subsequently. Students using SemNet® tend to develop habits such as looking back and planning ahead.

In thinking about the development of critical thinking, Kuhn (1999) describes three kinds of meta-knowing: metacognition (described above), metastrategic skill (the ability to apply consistent standards to the evaluation of ideas across space and time), and epistemological meta-knowing (in which knowing is understood as a process that entails judgement, evaluation and argument, and where there is the disposition as well as the skill to think critically). The latter is an important distinction because, since evaluative thinking entails work and effort, an individual must value the process in order to invest in it. Gorodetsky and Fisher (1996) obtained evidence of metacognitive gains among students using SemNet®.

EFFECTS WITH AND EFFECTS OF

Perkins (1993, p.89) makes a distinction between *effects with* information processing technologies and *effects of* information processing technologies. The *effects with* include amplifications of the user's cognitive powers during the use of a technology, as where the computer screen holds your ideas in place for you, allowing you to reflect on them The *effects of* are the spinoff effects that occur after users have finished using the technology, such as the ways in which the experience of having used SemNet® alters your thought processes and leads you to follow certain patterns in generating and organizing your ideas.

As an example of one of many after-effects, I have seen a student who was using SemNet® to study biology also plan her entire wedding in the form of a semantic network sketched on napkins in a restaurant. In another restaurant at another time, a professor was reviewing an important hierarchy in music: as he recited it, you could see him visualizing the images in his SemNet® screens. When individuals engage in semantic networking as a knowledge analysis process, the evidence suggests that they benefit from both the *effects with* and the *effects of* this technology.

I SEE WHAT YOU ARE THINKING!

Students in middle school are just beginning to engage in formal thinking. Wouldn't it make sense to make those formal thinking skills as clear, as explicit, and as visible as possible? Jay, Alldredge and Peters (1990), studying the use of SemNet® in seventh grade biology, concluded that seventh grade students a) clearly distinguish between building and refining networks; b) began to develop metacognitive skills with respect to how they structure domain specific information; c) identified and addressed discrepancies in their underlying knowledge by developing and using specific relations; and d) found it easy to use SemNet® for organizing and reorganizing their ideas. The researchers felt that SemNet® had distinct advantages over noncomputer aided mapping.

When instructors construct semantic networks and use them in their teaching, they convey their mental models of their areas of expertise to their students in as explicit and clear a manner as they possibly can. Whether they use semantic networks as lecture tools, as a medium for teaching across the internet, or as study tools, they are explicitly sharing not only their knowledge with their students but also the ways in which they organize that knowledge.

In 1990, Bradford, Gittings, Merkin and Morgan added video images to SemNet® semantic networks using a tedious, manual process. The authors felt that the SemNet®, by linking mapped meaning with video images, could transform video into a rich source for learning. A similar experiment with SemNet® as a hypermedia tool was done by Brigham, Hendricks, Kutcka, and Schuette (1994). It took just seven years from the first feasibility study of embedding video in semantic networks to robust practice. Professor Jack Logan (1999) uses semantic networks to teach two popular music courses via the internet. His nets contain a total of about 20,000 SemNet® graphic frames with links to about 1000 video clips, pictures and musical pieces. Production of such a course is facilitated by several versions of SemNet® that automatically convert SemNet® graphic frames into colored HTML clickable images with embedded links. Logan also has created hypertext links between key ideas in the semantic networks and the same ideas in his on-line text.

BROADCASTING TO THE SUBCONSCIOUS

Baars (1988) has spent his life studying the subconscious. He sees the conscious mind as the tiny tip of the iceberg, resting on and supported by a vast array of subconscious modules which work more or less independently from one another and in parallel. This is in fact a widely shared model of the mind. One module may be trying to match a name to a face you saw yesterday, for example, while another struggles to solve a math problem you were working on before lunch, and yet another is thinking about your date tonight. These subconscious modules account for the solutions that pop out of your head when you put things "on the back burner." According to Baars (1988), each concept pair that is brought into working memory and joined by a relation is broadcast to every subconscious module in the mind. This is one of the more potent *"effects of"* benefits derived from constructing a semantic network as described by Perkins (1993) above. As students construct semantic networks, they are making data available to the various modules in the mind involved in learning biology. It is also consistent with Pinker's (1994) theory that there are numerous innate, hard-wired modules in the brain for assimilating certain kinds of information. It is likely that Baars' modules and Pinker's modules are one and the same, viewed from psychology and linguistics respectively.

CREATIVITY VS SPECIFICITY

Buzon and Buzon (1993) believe that mind-mapping stimulates creativity. I suspect that this creativity, when realized, is promoted by keeping the links in Mind Maps unlabeled. Open links add fluidity and flexibility to one's thinking. With SemNet® as a learning tool in biology, we have a different aim: to add *precision* and *systematicity* to one's thinking. This is achieved largely through the use of labeled, bidirectional links.

COOPERATIVE NET-BUILDING

Students can construct networks working collaboratively in groups of two, three or four students. Group work is valuable in many ways, including the dialog it generates, the peer tutoring that often occurs, the overall sharing of skills and knowledge among students, and the community of thinkers and learners that develops (Johnson, 1974; Johnson & Johnson, 1975/1991; 1983; Johnson, Johnson, & Smith, 1991; Slavin, 1983). Groups that work well together are like well-oiled machines – they work in harmony and create beautiful nets. Peer evaluations are essential in monitoring group progress and evaluating each individual's contribution to the group. Students are asked to evaluate themselves as well as others in their group regarding their contributions to the creation of a semantic network. Dimensions of evaluation include students' biology knowledge, their leadership in the group, their net-building skills, and their computer skills. The consensus among peers about who did the most work and who made the most valuable contributions is usually very high. Student grades are raised or lowered relative to the group grade on the basis of such evaluations.

SEMNET®-BASED ASSESSMENT

Students can assess their knowledge in part with SemNet® frames that have the central concept masked. The task is to identify the masked concept based upon its links to related concepts. Many students say such questions make them "think differently." The items require students to evaluate more information about each concept than do standard multiple choice questions. At the other extreme, I often test students by asking each one to construct a semantic network about a topic. There are many possible ways to assess learning with SemNet® along the continuum between these two poles. For a more extended discussion of the SemNet® as an assessment tool, see Fisher (in press a).

EVALUATING A STUDENT NETWORK

SemNet® knowledge construction exercises can be specific or general. Some exercises are specific in that they are designed to build such cognitive skills as linking ideas into categories, building hierarchies, differentiating between big ideas and smaller ones, creating temporal flows, constructing causal chains, or naming distinguishing characteristics of a concept. More general assignments take the form of "Describe every thing you know about a cell (or photosynthesis, or evolution)." Regardless of the assignment, the four Cs are useful guides to evaluation: *completeness*, *coherence*, *correctness*, and *conciseness*. When a student constructs an entire network about a topic, it is typically judged qualitatively as if it were an essay. Such nets provide unusually clear views of how a particular student or student group was thinking at a point in time. There is no single right or wrong answer when it comes to a network of ideas. Rather, there is a range of quality from exquisite to dismal.

Evaluating nets is a task appropriate for students as well as the instructor. In fact, students learn a lot from reviewing other students' nets. They get insights into how to organize their ideas more effectively as well as things to clearly avoid in constructing a network, and they learn that different student groups can tackle the same ideas from quite different directions.

RESOURCE DATA BANK

A semantic network can serve as an encyclopedia of ideas. Some examples of semantic networks as content resources include (a) display of a medical taxonomy (Komorowski, Aiken, Greenes, & Pattison-Gordon, 1987); (b) graphic representation of large knowledge bases (Fairchild, Poltrock, & Furnas, 1986; Furnas, 1986); (c) visualization of the world wide web (Fowler, Fowler, & Williams, 1997); and (d) display of a 50,000 word dictionary called WordNet (Miller, Beckwith, Fellbaum, Gross & Miller, 1990). These projects are each supported by their own software on main-frame machines, although WordNet can be downloaded and run on a PC as well.

A semantic network provides one of the few dictionaries in which you can look up a word whose name you are unable to remember. You simply look up a closely related idea and then look around that portion of the network until you recognize the word you are seeking.

OBTAINING THE TOOL

The SemNet® software (version 1.1 ß14c), the SemNet® User Guide, and sample biology and family nets are available at: http://www.BiologyLessons.sdsu.edu/

SUMMARY

Quillian, in studying a memory phenomenon known as spreading activation, showed us how to capture human semantic network and processing in a computer. His semantic network theory has been confirmed by an enormous body of research on semantic networks in such fields as cognitive science, psychology and artificial intelligence. The SemNet® software draws upon this combination of cognitive theory and computer technology to change the ways in which students engage in biology learning. SemNet® is a tool that can support students in their transition from superficial rote learning habits to deeper strategies for meaningful understanding. And it is also a tool that can help teachers attain a reasonable balance between hands-on learning and minds-on learning in their classrooms.

Semantic networks can capture both the learner's prior knowledge (structural knowledge – Jonassen & Wang, 1993) and the acquisition of new knowledge including processes of elaboration and conceptual change – as illustrated with concept mapping in West & Pines (1985). Semantic networks of an individual's declarative knowledge about a domain can be used to accurately predict that individual's ability to perform in that domain.

The interconnectivity of a semantic network provides an enormous source of power for information processing and retrieval in both human memory and computer-based semantic networks. Organizing one's knowledge effectively involves both cognitive and metacognitive skills whose acquisition and ability to transfer to new domains are supported by SemNet®. Current versions of SemNet® are primitive compared to what could be, but it still seems to offer the most powerful knowledge representation system presently available for microcomputers.

The academy in-group has yet to discover the power of knowledge mapping as a learning tool. The situation is similar to those described by Kuhn (1970) for scientific revolutions and McNeill and Frieberger for fuzzy logic (1993), such that funding for research and development in knowledge mapping is difficult to obtain. The great emphasis today is on group learning and the impact of discourse. Individual cognitions still exist, however, and play a powerful role in learning. Being able to make these cognitions tangible in a computer-based knowledge arena is an effective learning and diagnostic aid. The growing number of mapping strategies appearing on the internet attests to their practical utility.

I am confident that the day will come when knowledge mapping is embraced not only by the academy but also by the world. Further, as educated persons gain the kind of fluency in knowledge mapping that they now have in word processing, their ways of thinking will change in positive and productive directions and the world of knowledge mapping will advance profoundly.

It is important to note that not all forms of knowledge mapping are created equal. Those that more closely mirror how our brain handles information have a competitive advantage in meshing external representations (knowledge maps) to internal representations (in the mind).

While social activities and context can indeed influence learning, an individual's mastery of science remains a personal, idiosyncratic and self-regulated constructive process. Any thinking tool that can bring a science learner's latent and unclarified understandings to the fore in schematic form so they can be examined and challenged has the potential to improve both scientific literacy and scientific thought. Before learners can truly SHARE what they KNOW with others, each one first needs to become AWARE of what s/he KNOWS!

Knowledge representation, vital from ancient times, has grown in sophistication from the wall of the cave to the video display of the desktop computer – and has increasingly empowered its users along the way. As my colleague and coauthor Jim Wandersee says, today's complex memory and processing tasks for doing science require cognitive, silicon prosthetics! Individuals benefit from this extended capacity just as groups do.

DAVID E. MOODY

CHAPTER 10

The Paradox of the Textbook

Conflicting Values

Paul Taylor was hired in the spring to become a first-year biology teacher at a small private school the following fall. The school's policies permitted Paul considerable latitude in the selection of a textbook for the course, and he spent the first half of the summer immersed in a careful review of all the textbooks on the market. After much deliberation, he selected a text that he felt embodied the most enlightened approach to instruction in biology: an approach designed to engage students in activities that brought them into direct contact with living things and with the materials and events that support life processes. The text was also designed to be highly accessible (that is, easy to read) for ninth or tenth-grade students. If this meant that the text was not encyclopedic in coverage or dictionary-like in conveying technical terms, those were sacrifices Paul was prepared to make.

Among the parents of the students in Paul's class that fall were two doctors, a psychiatrist and a neurologist, both of whom were dedicated to securing for their children the highest possible quality education in biology. Shortly after the school year began, one of these physicians contacted the Director of the school and asked a series of pointed questions regarding Paul, his qualifications to teach biology, and the textbook he had selected for the course. In the weeks and months that followed, the two doctors mounted a sustained campaign to change the basic thrust of the way Paul was teaching biology, with particular attention to the textbook he had assigned. They demanded a course that was far more technical, abstract, and conventional than the one Paul had constructed. They brought their concerns to Paul directly in a long series of meetings, as well as to the Director of the school and to the school board. In the end, the school administration supported Paul to the extent that he was allowed to continue using the text he had selected. As a kind of compromise, however, the school also ordered an additional textbook of a more conventional variety, to be made available for supplementary reading for students who desired it.

As a first-year teacher, Paul was extremely demoralized by the length and intensity of the battle in which he found himself engaged. At the end of the year he resigned his position and never returned to the classroom.

INTRODUCTION

It would be difficult to overstate the significance of the textbook as a determining factor in the science curriculum generally, and in biology in particular. Within the parameters laid down by the text, individual teachers may have broad discretion to exercise independent judgment in the design and implementation of their courses; but whether they choose to exercise such discretion is another matter. Survey after survey confirms the extent to which teachers express confidence in the textbooks they assign for their students (Weiss 1978, 1987; Yore, 1991). The corollary of this confidence is that the text, for most teachers most of the time, functions as a virtual blueprint for the curriculum (Elliott and Woodward, 1990; Gottfried and Kyle, 1992). So intimate is this relationship that the present chapter will at times consider the textbook and the curriculum as all but interchangeable elements of the educational system.

Textbooks are often said to be much maligned but little studied (Walker, 1981; Yore, 1991); but in recent years that situation has begun to be rectified. Research now exists that begins to reflect the manner in which textbooks actually function in science classrooms. We have begun to understand not only the extent to which teachers rely upon textbooks, but also the reasons why they rely upon textbooks, and the attitudes that accompany those reasons. The nature of the student's interaction with the printed page has also come under scrutiny, including the role that graphic representation tools may play in enhancing the student's relationship with the text.

The present chapter examines the biology textbook from multiple perspectives. The first of these consists of a selection of the findings of empirical research. The second perspective is a historical one, as we review the development of the biology textbook itself. The third section consists of a contemporary cross-section of opinions and conclusions about science texts. Together, these three ways of approaching and understanding the textbook provide the necessary background to appreciate the potential contribution of graphic representation tools for accessing meaning from textbooks. This approach is congruent with the constructivist learning theory that underlies this volume as a whole. Examining the textbook from the three perspectives described is designed to impart some of the contextual factors and other elaborations (Lloyd, 1990) that are needed to give the material life and relevance.

This examination is designed to bring into focus a fundamental paradox in the role that textbooks play in the biology classroom. As suggested in the foregoing vignette, a deep chasm separates what many thoughtful observers believe textbooks should be from what they are; a similar gulf separates how biology teachers actually employ textbooks from the way many believe they should employ them. The graphic representation tools described in this volume can play a crucial role in bridging these gaps, and so in assisting teachers such as Paul Taylor. In this way the textbook can function, not as a saint or a sinner, but as a solid citizen in the educational enterprise.

THE FINDINGS OF EDUCATIONAL RESEARCH

The present section reviews a series of interrelated studies pertaining to the use of the textbook in science instruction. The first is a trio of studies regarding the efficacy of graphic representation techniques as tools for extracting meaning from instruction. These findings stand in contrast to a second group of studies describing the manner in which teachers actually employ textbooks most of the time. This contrast and its implications raise the issue of the possible efficacy of concept-mapping as a tool for teachers as they think about the curriculum. The review concludes by showing the manner in which the foregoing findings fit within the broad framework of constructivist learning theory.

Okebukola (1990) and others have demonstrated the efficacy of concept mapping as a tool to facilitate learning of difficult concepts in biology, specifically in genetics and ecology. In his study, high-school-age students in Nigeria were taught the technique of concept mapping according to the six steps enumerated by Ault (1985). The students were given the opportunity to employ these techniques in their efforts to understand three-week units in genetics and ecology, while a control group spent a comparable amount of time engaged in more conventional study techniques. The concept-mapping group performed significantly better than did the controls on a subsequent test of meaningful learning in the two fields.

In Okebukola's study, the activity of concept mapping was employed as an immediate adjunct to instruction. In a pair of mutually complementary studies, concept mapping has also been shown to facilitate learning when constructed both before a unit of instruction, as an advance organizer, or when constructed following the unit, as a so-called "postorganizer". Willerman and MacHarg (1991) studied concept mapping as an advance organizer for eighth-grade students enrolled in a life science class and found significant effects upon subsequent learning. Spiegel and Barufaldi (1994) employed a similar exercise with students of college age, but here the concept-mapping activity occurred following the students' exposure to the material. Even in this case, the discipline of arranging concepts according to their hierarchical and other relationships resulted in an improvement in understanding and retention. Taken collectively, the implications of these three studies suggest that the subject matter of biology is so conducive to facilitation through the concept-mapping technique that it proves efficacious at every point in the learning sequence.

Spiegel and Barufaldi's study focused attention specifically upon learning directly from passages of text, and so warrants closer examination in the present context. In their study, community college students who were enrolled in a course in anatomy and physiology served as subjects. The students were asked to read short passages of text (approximately 200 words) and were tested on their recall. An experimental group received sixteen hours of instruction in study skills, including in the construction of graphic postorganizers, while a control group relied upon whatever study skills they brought to the experimental situation. According to Wandersee (1988), such skills are usually limited in their nature and variety.

Also of interest to the investigators were the effects upon recall and retention of instructing subjects in identifying various structural characteristics of reading passages: cause and effect, classification, enumeration, generalization, and sequence. Such instruction, they discovered, did not in itself result in any measurable increase in learning from textual material. Only when coupled with the students' active construction of graphic representations did training in the recognition of text structure produce a significant improvement in posttest scores. These findings suggest that metacognitive strategies employed in concert may have synergistic effects upon student learning.

Although the efficacy of concept mapping as a tool for helping students understand science content has been demonstrated by research, most teachers continue to employ their textbooks in a more conventional and prosaic manner. According to survey questionnaires and interviews, teachers at both the elementary and secondary levels imbue their textbooks with almost unquestioned authority, and they assign reading passages to their students in the somewhat naive expectation that the text is generally comprehensible (Yore, 1991; Shymansky, Yore & Good, 1991). DiGisi and Willett (1995) found that teachers at every level of high-school biology instruction — basic, college prep, honors, and advanced placement — rely upon their textbooks for multiple purposes, including reinforcing instruction given in class, as well as to introduce and convey new material entirely. At neither elementary nor secondary levels do teachers generally appreciate the nature or the difficulty of the metacognitive skills required for effective learning from texts in this manner. Perhaps for this reason, few teachers employ concept mapping or other comparable cognitive techniques in an effort to help their students make sense of their texts.

These data regarding teachers' relationships to textbooks and the subject matter of biology suggest the utility of preparing teachers to grapple directly with the complexities of the curriculum. In so doing, they are beginning to think for themselves about issues that are ordinarily left unquestioned by virtue of the authority of the text. Graphic representation tools may prove especially appropriate in this context, as suggested by a pair of studies that describe direct encounters with the curriculum for teachers at opposite ends of the life science curriculum. Starr and Krajcik (1990) found that sixth-grade teachers exhibited a marked evolution in the clarity, sophistication, and complexity of their maps over time. The element of improvement over time reappeared in the work of Edmondson (1995) who studied the evolution of curricular understanding as enhanced by concept mapping among teachers at the veterinary school at Cornell University. Here too the successive articulations not only brought clarity to the subject matter, but also served to facilitate the resolution of conflicts and contradictions in the views held about curriculum among the members of the faculty. Edmondson felt the exercise of concept mapping was particularly helpful in facilitating case-based and interdisciplinary approaches to subject matter.

The overall situation suggested by the findings of research tends to suggest the limitations of the distinction between learning science through experience as opposed to learning through reading. The actual nature of effective, meaningful learning in science is much more fluid and dynamic than is suggested by any easy, either-or dichotomy. In fact, both learning through experience and learning through reading have their place in an effective program of instruction. More to the point, both can be understood in terms of the crucial role played by the student's own construction of meaning from the materials available to him or her. Whether in the laboratory or in the encounter with the printed page, the student's prior knowledge plays a major role in mediating the interaction with the learning activity. Similarly, the student's independent activity of forging relationships, discerning distinctions, establishing continuities, and all the other elements of "minds-on" learning contribute to a successful outcome in reading as much as in any other learning activity (Holliday, Yore, and Alvermann, 1994; Lloyd, 1990).

BIOLOGY TEXTBOOKS: HISTORICAL OVERVIEW

The significance of the text is evident not only in the degree of teachers' reliance upon them, but also in the public controversies to which textbooks often give rise. Within the domain of biology, disputes over the content of texts have been especially virulent. These disputes occur repeatedly with respect to the incorporation and treatment of evolution, and with respect to other topics as well. The history of these disputes is instructive in understanding the forces that shape the composition and development of textbooks. By understanding the broad historical and sociological pressures at work, we can better see the nature of the finished product. Such an understanding is the necessary foundation for a full appreciation of the analysis of textual material by means of graphic representations.

Christy (1936) conducted the first systematic study of the nature of secondary-level biology instruction as it unfolded during the nineteenth century. His findings, however, were confined to an unpublished doctoral dissertation, and so have not been widely reported in the succeeding literature. A few years later, however, Cretzinger (1941) examined fifty-four textbooks, each intended for use at the secondary level, published in the United States between 1800 and 1933. Cretzinger understood that what he was examining represented

> ... fundamentally a history of what was taught in the field of biological science in the secondary schools of the United States ... it being generally understood that the curriculum of any field in the past was largely what was found in the textbooks of that time. (Cretzinger, 1941, p. 311)

Cretzinger performed a content analysis of each of the texts under examination. He established eight categories that embraced all the material in the texts. These included cell theory, the germ theory of disease, and theories on the origin of life; the concepts of anatomy, health, and heredity; principles of taxonomy; and the theory of evolution.

Cretzinger found that biology did not emerge as a distinct field of study at the secondary level until after 1900. Throughout the nineteenth century, textbooks were in the fields of natural history, botany, zoology, and physiology. The unified field of biological science that emerged after 1900 was portrayed in a cautious and conservative manner in secondary level texts, as "writers showed little interest in scientific theories or discoveries that opposed deep-seated, established opinions or superstitious beliefs."

A somewhat more focused examination was conducted by Hellman (1965) who was interested specifically in the treatment of evolution in textbooks. As Cretzinger had done, Hellman noted that a time lag occurs between developments in science and their corresponding reflection in texts, "partly because it takes some years to prepare a textbook manuscript, and partly because the theory must undergo a period of opposition and be tested by the rigors of debate." Nevertheless, Hellman found the first treatment of evolution in a text composed as early as 1888, H. Alleyne Nicholson's *Introduction to the Study of Biology* (as cited in Hellman, 1965). An influential text by Sedgwick and Wilson, *An Introduction to General Biology*, also contains a brief reference to evolution in its second (1890) edition (as cited in

Hellman, 1965). Sedgwick and Wilson's volume was more notable, however, for a new organizational structure, away from one based upon taxonomy, in favor of a systematic examination of the principles on which all life processes are based.

By 1926, according to Hellman, the "principles" approach was well-established in secondary level texts, and, in at least one text, evolution was considered fundamental to this approach. In Holmes' *Introduction to General Biology*, "evolution was no longer being considered a controversial theory or an hypothesis, but a unifying law which gave biology a meaning" (as cited in Hellman, 1965, p. 779). Hellman concludes that, notwithstanding an initial period of hesitancy, writers of secondary-level biology texts eventually did begin to incorporate evolution correctly, and he takes no notice of any sustained or unified forces of opposition to it. In this respect, Hellman's article may be considered as the calm before the storm, since in the years immediately ahead, the entire discussion of biology textbooks was largely, though not entirely, dominated by the issue of evolution.

A more contentious tone appeared in Troost (1968), who felt strongly that evolution was not accurately represented. A shot across the bow of the forces of opposition to evolution is apparent in his opening words:

> From the very beginning of biological education in America, it was clear that organic evolution was suppressed. That the teaching of evolution was historically only a trivial aspect of biological education is a conclusion which follows from an analysis of textbook content, courses of study, and curriculum committee recommendations. (Troost, 1968, p. 300)

Troost proceeds to make his case largely on the basis of the findings of Cretzinger (1941) and Hellman (1965), though he marshalls evidence from other sources as well. In Troost's account, no allowances are made for any time lag between scientific discovery and the incorporation of new ideas in texts. Instead, the belated appearance of evolution was in his view an act of deliberate oversight, and one which has never been fully corrected. Troost did not attempt to describe or to characterize the basis for the opposition to evolution. He took note, however, of the new trend represented by the BSCS generation of texts that appeared during the 1960s, and he urged that still more be done. In fact, he proposed that the time had arrived for evolution to become in biology classes what Einstein's general theory of relativity is in physics classes.

Grabiner and Miller (1974) also seized upon the issue of evolution in their review of secondary-level biology textbooks. Specifically, they asked, what were the effects upon textbook coverage of the Scopes trial of 1926? Most onlookers had considered the trial a moral victory for the forces of science; Grabiner and Miller found the actual effects were of another nature. The text John Scopes himself employed in his classroom in Tennessee, George William Hunter's *A Civic Biology* (as cited in Grabiner and Miller, 1974) made a brief but honorable mention of evolution prior to the trial. However, its coverage of evolution was curtailed in editions published in the aftermath of the trial. This result was emblematic of what Grabiner and Miller found in case after case: a reasonably fair and complete presentation of evolution in textbooks published before the trial, and a marked diminution subsequently. The authors conclude,

The evolutionists of the 1920's believed they had won a great victory in the Scopes trial.
But as far as teaching biology in the high schools was concerned, they had not won; they
had lost. Not only did they lose, but they did not even know they had lost. . . . That the
textbooks could have downgraded their treatment of evolution with almost nobody
noticing is the greatest tragedy of all. (Grabiner and Miller, 1974, p. 836–837)

The treatment of the issue of secondary-level biology textbooks underwent a quantum leap in breadth of examination with the appearance in 1977 of Nelkin's *Science Textbook Controversies and the Politics of Equal Time*. Nelkin's booklength treatment closely followed the eruption of intense controversy in a number of states. She describes in detail the "space-age fundamentalism" for which evolution in texts represents a major threat.

However, the social sources of textbook controversies draw upon wider pools of discontent as well. These include a generalized sense of disillusion with the effects of science and technology and a perversely democratic ideology, according to which all ideas are created equal and therefore warrant equal time in the politics of the curriculum.

Skoog (1979, 1984) gave the issue an even more definitive treatment with his highly quantitative approach. Skoog examined over a hundred secondary-level textbooks published in the United States since 1900 and subjected their contents to careful word-counts of topics associated with evolution. His hard-data approach yielded the fairly unequivocal conclusion that evolution had been treated "in a consistently cursory and non-controversial fashion." Skoog was careful to measure the sharp increase in attention to evolution contained in the BSCS series of texts, but even these witnessed a decline in coverage following the heady days of their youth in the 1960s.

A somewhat more detached view of events was presented by Rosenthal and Bybee (1986), who examined not only biology textbooks but the entire course of development of biology as a subject in the curriculum. For these authors, the controversies swirling around evolution were only part of a larger drama. They found that from its inception, the concept of a general biology course for secondary level students has been riven by alternative visions of its basic intent. For one group, the aim has been to present biology as a mature scientific discipline, as the study of life; for the other, the aim has been to select only those themes from the larger discipline of most interest and value to the adolescent. This drama, they found, both underlies and embraces the more specific controversies surrounding evolution, sex education, ecological concerns, and other issues with a more topical focus. Rosenthal and Bybee (1986) noted that as early as 1909,

Topics such as the sources and biological importance of food, the relation of organisms
to food production and food destruction, hygiene of food preparation and digestion,
sanitation, the effects of alcohol and narcotics, the risks involved in patent medicines, the
role of living organisms in producing clothing and building materials, conservation,
disease, public health, and sex were advocated as proper subjects for a biology course.
(Rosenthal and Bybee, 1986, p. 137)

My research (Moody, 1996) examined the structure of the subject matter of biology in eight texts published during the 1980s (Figure 10.1) and 1990s (Figure

10.2). The results were conveyed in graphic representations not dissimilar to orthodox concept maps, although the high level of abstraction in these maps mitigated against the inclusion of informative phrases on the lines connecting terms.

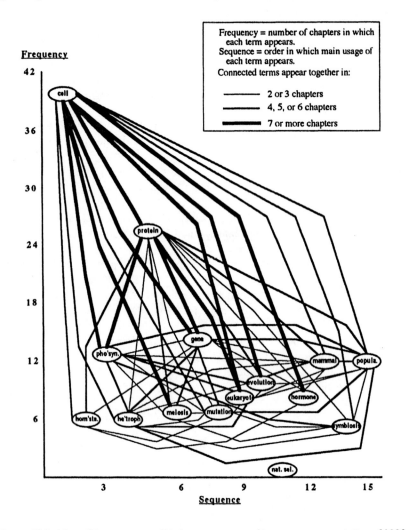

Figure 10.1. Map of the structure of biology as presented in a text representative of 1980's generation of textbooks, Modern Biology *by Otto & Towle (1985). The graphic was first presented in Moody, 1996. The order of the terms relative to the 'x' axis represents the sequence in which the main usage of each term occurs throughout the text. The height of each term relative to the 'y' axis represents the number of chapters in which the term appears. The lines connecting terms vary in width according to the number of chapters in which the connected terms both appear.*

The maps indicate that the treatment of evolution in texts published during the 1980s was not compatible with the significance of the topic of evolution for the discipline of biology. In the generation of texts published during the 1990s, however, a marked shift of emphasis occurred in which evolution was elevated to a much more significant position.

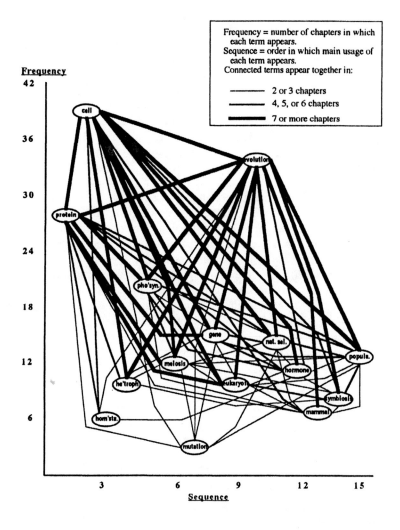

Figure 10.2. Map of the structure of biology as presented in a text representative of 1990's generation of textbooks, Modern Biology *by Towle (1993). The graphic as first appeared in Moody, 1996. For explanation, see legend for Figure 10.1.*

The structure of the subject matter was assessed by designating key terms as representative of the major topics of biology, and monitoring the occurrence and co-occurrence of those terms as they were deployed throughout the texts examined. The study measured the

(1) *frequency* (number of chapters) in which each term appeared;

(2) *sequence* in which the main usage of each term appeared; and

(3) mutual *proximity* of use of each pair of terms.

These results are conveyed collectively in graphic representations such as those in Figures 10.1 and 10.2. In those figures, the order of the terms relative to the 'x' axis represents the sequence in which the main usage of each term occurs throughout the text. The height of each term relative to the 'y' axis represents the number of chapters in which the term appears. The lines connecting terms vary in width according to the number of chapters in which the connected terms both appear. The simultaneous depiction of sequence, frequency, and proximity of use of selected terms was designed to reveal the skeletal structure of each textbook as a whole.

In none of the textbooks examined was evolution the leading, guiding, or overarching concept in the structure of the subject matter; that honor went almost exclusively to the concept of "cell". The terms "gene" and "protein" were also often prominent in the structure of the subject matter. The terms associated with population biology ("population" and "symbiosis") tended to occur in subordinate positions in the textbook structures. Natural selection, the lynchpin of evolution, was generally relegated to a similar role. An analytical, reductionist approach to the material was the structure clearly preferred in most secondary-level biology textbooks.

By examining the history of the biology textbook as it has developed in the United States, it is possible to see that there exists no easy or obvious relationship between developments in the discipline of biology itself and the secondary-level text. Rather, the textbook is the result of a more complex network of forces, which must be understood to grasp the nature of the finished product. Some of these forces are examined in more detail in the section that follows.

A CROSS-SECTION OF CONTEMPORARY VIEWS

With the foregoing historical sketch of the development of textbooks as background material, we turn now to a cross-section of contemporary views regarding the nature, value, and utility of the secondary-level biology textbook. This cross-section, in turn, will pave the way for a succeeding section, in which we examine more specifically the ways in which graphic representations may be used to elucidate the characteristic content of biology textbooks.

Eylon and Linn (1988) conducted a comprehensive review of the state of the research literature pertaining to science education as a whole, and reflected upon the implications of this literature for the reality of life in the classroom. They found that in science generally, not only in biology, the composition of textbooks must bear a significant share of the burden for the inferior quality of the finished product:

The table of contents of most precollege texts reads like the course catalog for a four-year college. These books provide fleeting coverage of numerous topics rather than integrated coverage of central topics. The new vocabulary in a one-week science unit often exceeds that for a one-week unit in a foreign language. (Eylon and Linn, 1988, p. 252)

A prevailing sentiment against textbooks in biology classrooms appeared in Gottfried and Kyle's (1992) study regarding approximations to the "biology education desired state." On the basis of a questionnaire, Gottfried and Kyle classified teachers according to whether their approaches to classroom instruction were Textbook Centered (TC) or Multiple Reference (MR). TC teachers were defined as those who relied exclusively upon a single textbook for curriculum planning and implementation. MR teachers relied on no textbook at all or on multiple sources for the same instructional purposes. A subsequent analysis revealed that the TC teachers were inferior to their MR cohorts with respect to whether their classrooms approximated the biology education desired state, as defined by Project Synthesis and the National Science Teachers Association. That state was held to consist of a complex mix of factors, including open-ended, problem-centered, flexible approaches to subject matter. Notwithstanding the possibility of a kind of circular reasoning underpinning the logic of this article, it serves to reinforce evidence of a widespread, highly critical attitude toward biology textbooks.

The critique set forth by the National Research Council (1990) in its review of biology education as a whole may serve to summarize and to emphasize the predominant outlook:

There is clearly a tension between the demands for textbook comprehensiveness and the limitations of textbook size. The usual casualty is the presentation of biology as an experimental science. In that respect, the books merely amplify the growing pressures of tests and curricula to de-emphasize the process of discovery and to portray biology as the worst kind of literature — all characters and no story.

In summary, current biology textbooks are an important part of the failed biology curriculum. They are often not selective in what they present and lack both a broad conceptual basis and a refined understanding of specific subjects. They emphasize memorization of technical terms. They have many misleading and superfluous illustrations. The books are different, but a tendency toward uniformity and mediocrity can be seen in recent years. (National Research Council, 1990, p. 30)

The widespread and deeply negative attitude toward textbooks must be viewed in the context of a countervailing fact of great significance: the overwhelming reliance upon and satisfaction with their texts expressed by the large majority of practicing biology teachers (Weiss, 1978, 1987; Yore, 1991). This paradoxical situation was addressed very directly by Roth and Anderson (1988), who described the anguish of a typical first-year teacher, one who has been taught from multiple sources that over reliance upon the textbook is a cardinal classroom sin. For such a teacher, the textbook may seem like an island of clarity in a confusing sea of more intractable issues, such as those pertaining to classroom management and the immediate school environment. Thus, as much as the textbook is under fire from one end of the spectrum, it remains warmly appreciated by and closely attached to those in the trenches.

By general consensus (Broudy, 1975; Zebrowski, 1983; National Research Council, 1990) the textbook publishing industry as a whole is an arena of intensely conflicting pressures, not necessarily guaranteed to engender the finest possible product. A fundamental discontinuity exists between the actual learning needs of the student and a myriad of other considerations. This disjunction begins with the deliberations of the teacher who is in fact the primary consumer of textbooks, in the sense that it is he or she who orders a given title from an array of competing alternatives. And the teacher's decision, as Broudy (1975) has emphasized, may or may not have much to do with the experiences of students as readers of the texts. Rather, as noted previously, the text often serves to define for the teacher the curriculum of the course: it prescribes the broad parameters of instruction, the order in which topics shall be introduced and the relative emphasis upon each, as well as an encyclopedic level of detail that the teacher can consult in any contingency.

Beyond the teacher's immediate needs, a host of additional factors compete to shape the composition of texts. These occur at several levels, including the concerns of parents and community groups, universities and professional organizations, and governing authorities at the local district, state, and federal levels. Textbook adoption committees in a handful of key states exert particularly great influence on the composition of texts.

As has been widely documented (Nelkin, 1977; Weinberg, 1978; Berra, 1990), textbook adoption committees established by state legislatures in Texas, California, and elsewhere exert an inordinate influence over the shape and composition of textbooks in every domain of K–12 education. Membership on these committees is more often determined by voting constituencies and special interests than it is of gathering together the highest possible level of professional expertise. While expert opinions are often consulted, time constraints, local politics, and other extraneous considerations often distort the decision-making processes of such committees. The decisions made, moreover, have a disproportionate impact upon the marketplace, since their effect is to eliminate some products altogether, and profoundly to shape the character of those that survive the cut. The available evidence indicates that this is not a state of affairs conducive to the development of lively, original, and readable textbooks. On the contrary, many texts appear to be designed with the aim of being inoffensive to the broadest possible constituency, so that the finished product is dilute, flat, tepid, without the spark of original thinking or an authentic authorial voice (Tyson-Bernstein, 1990).

GRAPHIC REPRESENTATIONS AS TOOLS FOR ACCESSING TEXTBOOKS

The foregoing analyses of the historical development of the secondary-level biology textbook, and a contemporary cross-section of views regarding the textbook today, may serve to suggest the critical function that concept maps and other graphic representations may play in enabling students to make use of their texts. The evidence assembled in this chapter suggests that the biology textbook in particular, however mild-mannered it may appear on the surface, is the outcome of a seething cauldron of conflicting forces and tendencies. From publisher to committee to teacher to public opinion, the textbook has been designed to serve purposes more or less extraneous to the act of reading by the student. Whether that extraneous pressure is exerted by large-scale, organized forces of opposition such as religious fundamentalists, or whether it is exerted by the fickle intent of a single acquisitions editor, the effect on the student is the same. The biology text has not been designed with the students' reading interests and abilities foremost in mind. On the contrary, it should be understood as a matter of course that some special devices or tools are virtual prerequisites to enable the large majority of students to extract meaning from their texts. We suggest that the knowledge representation techniques described in this book are among the best available tools of this kind.

Broadly speaking, there are four avenues through which graphic representations can assist in making textual material more accessible to students. The first of these occurs at the planning stage of the text, and involves the use of concept maps to clarify communication among the diverse parties involved in text preparation. The next level consists of representations that appear in the text itself, as conceptual aids introduced by the authors and designed to communicate directly with readers, whether students, teachers, or anyone else. The third level occurs within the walls of the classroom, and consists of maps and diagrams introduced by the teacher as advance organizers or other tools to facilitate student comprehension of their texts. The fourth level is the most concrete and immediate, and consists of the student's own construction of graphic representations as an aid in his or her direct encounter with the text.

As we have seen, the process of textbook development is one of close and often contentious interaction among a diversity of parties: author, publisher, adoption committees, teachers, community groups, and the upholders of the professional standards of the discipline. The sheer difficulty of finding clear ways of communicating about the intricate complexities of the curriculum often mitigates against the successful resolution of these disputes. This is the kind of difficulty, however, that the concept map is well adapted to circumvent. The strength of the concept map is that it shows the general pattern of relationships among a large number of concepts within the boundaries of a single page.

In recognition of the foregoing fact, concept maps have been appearing with increasing frequency within textbooks themselves. The successful deployment of such maps in this context, however, is by no means a foregone conclusion, since a degree of skill is required in the reading of the map itself. Unless this issue receives careful attention by the teacher, the concept map as presented in the text runs the risk of going the way of all the other material in the text. In particular, the hierarchical nature of relationships among terms is an issue that must be taught. As with any cognitive tool, issues that may appear obvious to the experienced user cannot be assumed on behalf of the novice. Notwithstanding these caveats, the introduction of concept maps into textbooks is an important innovation, and one whose consequences warrant the careful attention of educational research.

The next level of interaction in which graphic representations of textual material may be of benefit for student understanding pertains to the teacher's direct communications with his or her class. It is in this context that the circle diagram perhaps attains its greatest efficacy. By virtue of its inherent simplicity, the circle diagram is preeminently a teacher's tool, designed for quick communication and immediate comprehension. The teacher can design an appropriate circle diagram almost on the spot in order to convey the essential elements of an evening's reading assignment or the most fundamental characteristics of a biological process.

In the final analysis, no matter how good the text or how thorough the preparation by the teacher, the student must ultimately be left alone with the text. When that moment arrives, rudimentary reading skills are not likely to be sufficient to make reasonable sense of the subtleties of biological science for most students. At this moment the power and efficacy of SemNet® as a graphic representation tool reaches its apex.

Perhaps the most frustrating and intractable characteristic of biology as a discipline for the novice student is the great wealth of new vocabulary that the subject demands. Closely related to this characteristic is the intricate nature of relationships among biological terms: ultimately, everything is related to everything else. It is precisely this kind and quality of knowledge that SemNet® is uniquely designed to capture and represent. As Fisher (1990) has reported, in one exercise over two thousand concepts from a college course in biology were contained in a single network. While the high school student is unlikely to construct anything so elaborate, a more modestly interconnected series of frames may be well suited to making sense of a single chapter or topic.

So successful is SemNet® at enabling students to construct complex patterns of networks among concepts that it is possible to build nets in which one loses a sense of the territory as a whole.

Such a result is by no means inevitable, however, if the creator of the net desires a different result. To demonstrate this point, Figures 10.3, 10.4, and 10.5 represent three SemNet® frames designed to depict the broad scope of the material on human evolution in the 1990 edition of the BSCS Blue version text, *Biological Science: A Molecular Approach*.

The material occupies some 20 pages of text; it represents one chapter in a text with 26 chapters in all. In this short series of frames, it is possible to discern the boundaries and internal organization of the material in the chapter. In this demonstration we may also detect several of the general characteristics of SemNet® that enable it to facilitate learning so effectively:

1. Ease and economy of constructing representations: Because the underlying scaffolding and structure of representation are already in place, it is a relatively straightforward matter to focus on the cognitive elements of the task: Key concepts and the precise specification of relationships among them.
2. Emphasis upon interrelationships among parts: Fisher (1990) notes that the original inspiration for SemNet® arose from a classroom game in which students selected two terms at random from a large collection of biology concepts. The task was to describe how the two selected concepts are related. In a SemNet® network, the relational path connecting any two concepts is made explicit.

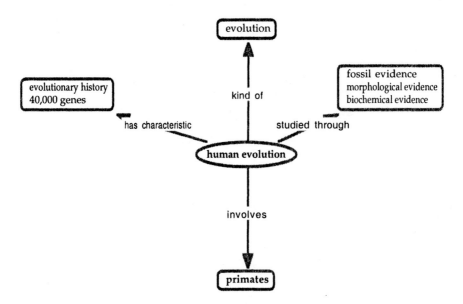

Figure 10.3. First of three SemNet® frames depicting the broad scope of chapter on human evolution in Biological Science: A Molecular Approach *(BSCS, 1990).*

3. Flexibility of use: In the construction of the present net, no particular effort was made to follow the precise sequence in which information was conveyed in the text, nor exhaustively to reproduce all such information. Rather, the emphasis was on organizing the concepts in a way that reflected the broad scope of the material.

It is important for the teacher to recall that sheer accuracy of the resultant representations is only one element to consider in evaluating students' work. As Fisher (1995) has emphasized, the benefit for the student lies as much in the process of generating constructions as in the shorthand knowledge that appears in the finished product. As a result, other important attributes to consider in evaluating SemNet® frames are (a) robustness; (b) connectedness; (c) functionality; (d) completeness; and (e) coherence, in addition to (f) correctness. To this list we might add the virtue of simplicity, as SemNet® helps students see at a glance the basic logic of a complex topic.

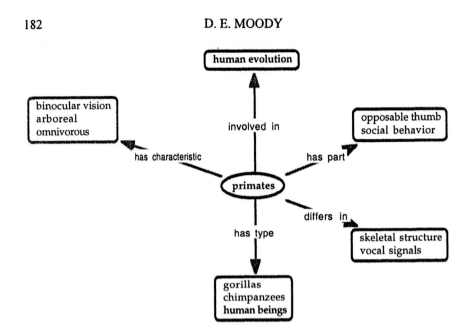

Figure 10.4. First of three SemNet® frames depicting the broad scope of chapter on human evolution in Biological Science: A Molecular Approach *(BSCS, 1990).*

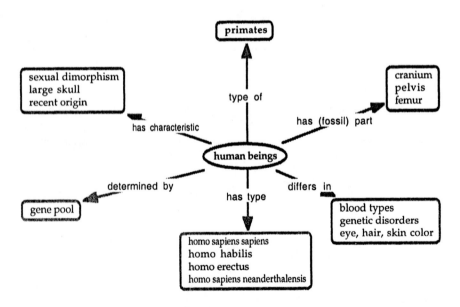

Figure 10.5 Second of three SemNet® frames depicting the broad scope of chapter on human evolution in Biological Science: A molecular Approach *(BSCS, 1990).*

The utility of graphic representations in making sense of biology knowledge is not limited to the illustrative channels described in this section. Concept maps, circle diagrams, and SemNet® may each be employed at each of the stages we have described. Their utility is dictated not by pre-existing rules, but only by the immediate needs and the imagination of the user. Ultimately, we may envision a community of teachers and learners with skilled access to each of these metacognitive tools, and to others as well. Indeed, their greatest utility may not emerge until such tools are so widespread as to appear commonplace, ordinary, and unremarkable.

CONCLUSION

The evidence assembled in this chapter tells a tale of a gradually increasing understanding of the processes that give rise to successful learning of science from the reading of text. It is no longer as accurate as it once was to say that textbooks are "much maligned but little studied." Although a wealth of fruitful avenues of research remain to be explored, it is not now the case that the textbook has been little studied. And to say that it has been maligned does not do justice to the objectivity and constructive spirit of much of the existing criticism. It would be fairer to say that the science textbook has been closely examined and hotly debated, with no clear resolution of the contested issues yet in sight.

As was suggested in the vignette with which this chapter begins, the secondary-level biology textbook in particular has been the focus of intense disputation. Among the generation of textbooks published during the 1990s, there are several whose presentation of evolutionary content is actually roughly commensurate with the significance of the topic in the field – a rare occurrence in the history of biology texts.

Other widely criticized tendencies in these texts, however, remain for the time being more intractable. In particular, textbooks continue to be laden with technical vocabulary and encyclopedic in scope rather than devoted to clarifying the essence of the subject matter. They tend to present the findings of science as a finished and polished body of knowledge, received from on high and imperishable, rather than as a rough-and-tumble arena of earnest inquiry. The net effect appears to be to pacify the innovative impulses of teachers and to stultify the interests of students, rather than to stimulate both student and teacher to be active consumers of biological knowledge and understanding.

The present chapter describes some of the ways in which the graphic representation of biology knowledge may serve to ameliorate this state of affairs. For reasons closely related to the nature of biology knowledge itself, the process of specifying networks of relationships among terms closely models the activity of cognitively constructing meaning that must take place for learning to occur, whether from a text or otherwise. As a result, such an activity proves useful before, during and after the student's direct encounter with the subject matter, and it may serve equally to help organize the thinking of teachers and others involved in curriculum development. As an adjunct to other instructional activities, graphic representations of textual material offer the biology teacher substantial benefits in exchange for a minimal investment of time and energy.

As fruitful and rewarding as recent research in this area has proven to be, several pressing empirical issues remain to be resolved, in addition to the more obvious practical concerns. The vagaries associated with the production and state-level adoption of textbooks are systemic in nature, and their resolution is a matter for policy-makers as well as the concerned citizen. What the research community can contribute to the fray are focused observations of such issues as the following: What is the effect on student achievement of alternative structures of the subject matter of biology? To what extent is effective learning from science text a product of appropriate prior knowledge and reading skills, as opposed to innate motivation and interest in the subject matter? What characteristics of textbooks are most conducive to generating positive affective responses as well as appropriate cognitive ones? The graphic representation tools described in the present volume are uniquely appropriate to contribute to the resolution of these issues. Some of the ways in which they may serve that function are explored in the chapters that follow.

Abrams, E. & Wandersee, J.H. (1995). 'How does biological knowledge grow? A study of life scientists' research practices', *Journal of Research in Science Teaching* (32), 649–663.

Aleixandre, M.P. (1996). 'Darwinian and Lamarkian models used by students and their representation', in K.M. Fisher & M.R. Kibby (eds.), *Knowledge acquisition, organization, and use in biology* (NATO ASI Series F, Vol. 148), New York, Springer Verlag, 65–77.

Al-Kunifed, A. & Wandersee, J.H. (1990). 'One hundred references related to concept mapping', *Journal of Research in Science Teaching* (27), 1069–1075.

Altman, L.K. (1997, December 9). 'Study on using magnets to treat pain surprises skeptics', *The New York Times* p. B11.

Ambron, J. (1988). 'Clustering: An interactive technique to enhance learning in biology', *Journal of College Science Teaching* (18), 122–144.

Amer, A.A. (1994). 'The effect of knowledge-map and underlining training on the reading comprehension of scientific texts', *English for Specific Purposes* 13(1), 35–45.

American Association for the Advancement of Science (1983*). Benchmarks for science literacy*, New York, Oxford University Press.

American Association for the Advancement of Science (1994, September 23). 'Vignettes: Shifts in biology' (Claus Emmeche & E. O. Wilson), *Science* p. 1901.

American Association for the Advancement of Science (1997, September 5). 'Vignette: Editing' (Cathy Yarbrough), *Science* p. 1445.

American Association for the Advancement of Science Project 2061 (1989). *Science for all Americans*, New York, Oxford University Press.

American Association for the Advancement of Science Project 2061 (1998). *Blueprints for reform*, New York, Oxford University Press.

Amlund, J.T., Gaffney, J. & Kulhavy, R.W. (1985). 'Map feature content and text recall of good and poor readers', *Journal of Reading Behavior* (17), 317–330.

Anderson, C.W., Sheldon, T.H. & Dubay, J. (1990). 'The effects of instruction on college non-majors' conceptions of respiration and photosynthesis', *Journal of Research in College Teaching* (27), 761–776.

Anderson, J.R. (1983). *The architecture of cognition*, Cambridge, MA, Harvard University Press.

Aristotle. (1952/1997). *Metaphysics* (R. Hope, Trans.), Ann Arbor, MI, The University of Michigan Press.

Arons, A.B. (1977). *The various language: An inquiry approach to the physical sciences*, New York, Oxford University Press.

Atrans, S. (1990). *Cognitive foundations of natural history*, Cambridge, UK, Cambridge University Press.

Ault, C.R. (1985). 'Concept mapping as a study strategy in earth science', *Journal of College Science Teaching* (15), 38–44.

Ausubel, D. P. (1963). *The psychology of meaningful verbal learning.* New York: Grune & Stratton.

Ausubel, D. P. (1968). *Educational psychology: A cognitive view.* San Francisco: Holt, Rinehart, & Winston.

Ausubel, D.P., Novak, J.D. & Hanesian, H. (1978). *Educational psychology: A cognitive view, 2nd ed.*, New York, Holt, Rinehart, & Winston.

Baars, B.J. (1988). *A cognitive theory of consciousness*, Cambridge, UK, Cambridge University Press.

Barinaga, M. (1997). 'New insights into how babies learn language', *Science* (277), 641.

Basili, P.A. & Sanford, J.P. (1991). 'Conceptual change strategies and cooperative group work in chemistry', *Journal of Research in Science Teaching* (28), 293–304.

Berra, T.M. (1990). *Evolution and the myth of creationism*, Stanford, CA, Stanford University Press.

Beyerbach, B. & Smith, J. (1990). 'Using a computerized concept mapping program to assess preservice teachers' thinking about effective teaching', *Journal of Research in Science Teaching* (27), 961–972.

Biological Sciences Curriculum Study—Blue Version (1990). *Biological science: A molecular approach, sixth edition*, Lexington, MA, D.C. Heath.

Bishop, B.A. & Anderson, C.W. (1990). 'Student conceptions of natural selection and its role in evolution', *Journal of Research in College Teaching* (27), 415–427.

Blakeslee, S. (1997, August 1). 'Study finds that baby talk means more than a coo', *The New York Times* p. A14.

Blakeslee, S. (1997, July 22). 'What controls blood flow? Blood', *The New York Times* pp. B7, B10.

Bourne, L.E.Jr., Dominowski, R.L. & Loftus, E.F. (1979). *Cognitive processes,* Englewood Cliffs, NJ, Prentice-Hall.

Brachman, R. J. & Levesque, H. J. (eds.). (1985).*Readings in knowledge representation.* Los Altos, CA: Morgan Kaufmann.

Brachman, R.J., Levesque, H.J. & Reiter, R. (eds.). (1991). 'Knowledge representation', *Special Issues of Artificial Intelligence, an International Journal* (49), available at http://www.elsevier.nl/inca/publications/store/5/2/1/1/4/0

Bradford, D., Gittings, E.C., Merkin, A. & Morgan, R.L. (1990). 'An exploration in hypervisual systems: Creating the video tab on a semantic network', paper submitted in the course Educational Technology 653, B. Allen, Instructor, Department of Educational Technology, San Diego, CA, San Diego State University.

Branaghan, R.J. (1990). 'Pathfinder networks and multidimensional spaces: Relative strengths in representing strong associates', in R. Schvaneveldt (ed.), *Pathfinder associative networks: Studies in knowledge organization,* Norwood, NJ, Ablex, 111–120.

Brigham, F.J., Hendricks, P.L., Kutcka, S.M. & Schuette, E.E. (1994). 'Hypermedia supports for student learning', paper presented at the annual meeting of the Indiana federation, council for exceptional Children, Indianapolis, IN.

Bronowski, J. & Mazlish, B. (1960). *The western intellectual tradition,* New York, Harper & Row.

Broudy, E. (1975). 'The trouble with textbooks', *Teachers College Record* (77), 13–35.

Brown, R. & McNeill, D. (1966). 'The "tip of the tongue" phenomenon', *Journal of Verbal Learning and Verbal Behavior* (5), 325–337.

Bruer, J.T. (1997). 'Education and the brain: A bridge too far', *Educational Researcher* 26(8), 4–16.

Brumby, M. (1979). 'Problems in learning the concept of natural selection', *Journal of Biological Education* (13), 119–122.

Brumby, M. (1982). 'Students' conceptions of the life concept', *Science Education* (66), 613–622.

Buzon, T. & Buzon, B. (1993). *The mind map book: How to use radiant thinking to maximize your brain's untapped potential,* New York, Plume Book (Penguin).

Cain, S.E. & Evans, J.M. (1979). *Sciencing: An involvement approach to elementary science methods,* Columbus, OH, Merrill.

Carey, S. (1987). *Conceptual change in childhood,* Cambridge, MA, MIT Press.

Champagne, A., Gunstone, R. & Klopfer, L. (1985). 'Naive knowledge and science learning', *Research in Science and Technological Education* (1), 173–183.

Chase, W.G. & Simon, H.A. (1973). 'The mind's eye in chess', in W.G. Chase (ed.), *Visual information processing: Proceedings,* New York, Academic Press, 215–281.

Chi, M.T.H., Feltovich, P.J., & Glaser, R. (1981). Categorization and representation of physics problems by experts and novices. *Cognitive Science,* 5, 121–151.

Chinn, C.A. & Brewer, W.F. (1993). 'The role of anomalous data in knowledge acquisition: A theoretical framework and implications for science instruction', *Review of Educational Research* (63), 1–49.

Chmielewski, T.L. & Dansereau, D.F. (1998). 'Enhancing the recall of text: Knowledge mapping training promotes implicit transfer', *Journal of Educational Psychology* (90), 407–413.

Chomsky, C. (1975). *Reflections on language,* New York, Pantheon.

Chomsky, C. & Commentators (1980). 'Rules and representations', *Behavioral and Brain Sciences* (3), 1–61.

Chomsky, C. (1988). *Language and the problems of knowledge. The Managua lectures,* Cambridge, MA, MIT Press.

Christianson, R.G. & Fisher, K.M. (1999). 'Comparison of student learning about diffusion and osmosis in constructivist and traditional classrooms', *International Journal of Science Education* (21), 687–698.

Christy, O.B. (1936). *The development of the teaching of general biology in the secondary schools* (Peabody Contribution to Education No. 21), Nashville, TN, George Peabody College for Teachers.

Clark, W.R. (1995). *At war within: The double-edged sword of immunity,* New York, Oxford University Press.

Clement, J. (1982a). 'Students' preconceptions in introductory mechanics', *American Journal of Physics* (50), 66–71.

Clement, J. (1982b). 'Spontaneous analogies in problem solving: The progressive construction of mental models', paper presented at the annual meeting of the American Educational Research Association, New York.

Clement, J. (1988). 'Observed methods for generating analogies in scientific problem solving', *Cognitive Science* (12), 563–586.

Clement, J. with Brown, D., Camp, C., Kudukey, J., Minstrell, J., Palmer, D., Schultz, K., Shimabukuro, J., Steinberg, M. & Veneman, V. (1987). 'Overcoming students' misconceptions in physics: The role of anchoring conceptions and analogical validity', in J. Novak (ed.), *Proceedings of the second international seminar: Misconceptions and educational strategies in science and mathematics, Vol. III,* Ithaca, NY, Cornell University Press, 84–97.

Clements, D.H. & Gullo, D.F. (1984). 'Effects of computer programming on young children's cognition', *Journal of Educational Psychology* (76), 1051–1058.

Cliburn, J.W.Jr. (1990). 'Concept maps to promote meaningful learning', *Journal of College Science Teaching* (19), 212–217.

Cobern, W. (1996). 'World view theory and conceptual change in science education', *Science Education* (80), 579–610.

Cohen, D.K. (1991). 'Revolution in one classroom (or then again, was it?)', *American Educator* 15(2), 16–23, 44–48.

Collette, A.T. & Chiappetta, E.L. (1994). *Science instruction in the middle and secondary schools, 3rd edition,* New York, Macmillan.

Collins, A. & Gentner, D. (1982). 'Constructing runnable mental models', in *Proceedings of the fourth annual conference of the cognitive science society,* Ann Arbor, MI.

Collins, A.M. & Loftus, E.F. (1975). 'A spreading activation theory of semantic processing', *Psychological Review* (82), 407–428.

Cooke, N.M. & McDonald, J.E. (1986). 'A formal methodology for acquiring and representing expert knowledge', *Proceedings of the IEEE* (74), 1422–1430.

Craig, R. (1997, December 8). 'Don't know much about history', *The New York Times* p. A23.

Cretzinger, J.I. (1941). 'An analysis of principles or generalities appearing in biology textbooks used in the secondary schools of the United States from 1800 to 1933', *Science Education* (25), 310–313.

Crovello, T.J. (1984). 'Computers in bioeducation: The expanding universe', *The American Biology Teacher* (46), 139–145.

Cushing, S. (1994). *Fatal words: Communication clashes and aircraft crashes,* Chicago, University of Chicago Press.

Dagher, Z.R. (1994). 'Does the use of analogies contribute to conceptual change?', *Science Education* (78), 601–614.

Dalton, B., Morocco, C.C., Mead, P.L.R. & Tivnan, T. (1997). 'Supported inquiry science: Teaching for conceptual change in urban and suburban science classrooms', *Journal of Learning Disabilities* (30), 670–684.

Davis, R., Shrobe, H. & Szolovits, P. (1993). 'What is a knowledge representation?', *AI Magazine* (14), 17–33.

Dawkins, R. (1986). *The Blind Watchmaker.* Harlow: Longman Scientific & Technical.

De Jong, O. & Brinkman, F. (1997). 'Teacher thinking and conceptual change in science and mathematics education', *European Journal of Teacher Education* (20), 121–124.

Demastes, S.S., Good, R.G. & Peebles, P. (1996). 'Patterns of conceptual change in evolution', *Journal of Research in Science Teaching* (33), 407–431.

Demastes, S.S., Settlage, J.Jr. & Good, R. (1995). 'Student conceptions of natural selection and its role in evolution: Cases of replication and comparison', *Journal of Research in Science Teaching* (32), 535–550.

Dennett, D.C. (1996). *Kinds of minds: Toward an understanding of consciousness,* New York, Basic Books.

DiGisi, L.L. & Willett, J.B. (1995). 'What high school biology teachers say about their textbook use: A descriptive study', *Journal of Research in Science Teaching* (32), 123–142.

Donaldson, M. (1978). *Children's minds,* New York, W. W. Norton.

Dorough, D.K. & Rye, J.A. (1997). 'Mapping for understanding', *The Science Teacher* (64), 37–41.

Driver, R., Squires, A., Rushworth, P. & Wood-Robinson, V. (1994). *Making sense of secondary science: Research into children's ideas,* New York, Routledge.

Dubin, R. & Taveggia, T.C. (1969). *The teaching-learning paradox,* Eugene, OR, Center for the Advanced Study of Education Administration.

Dunlap, D.W. (1997, November 22). 'Mission to put Manhattan on the map', *The New York Times* p. A24.

Edmondson, K.M. (1995). 'Concept mapping for the development of medical curricula', *Journal of Research in Science Teaching* (32), 777–793.

Elliott, D.L. & Woodward, A. (1990). 'Textbooks, curriculum, and school improvement', in D.L. Elliott & A. Woodward (eds.), *Textbooks and schooling in the United States,* Chicago, University of Chicago Press, 222–232.

Ericsson, K.A. & Simon, H.A. (1984). *Protocol analysis: Verbal reports as data,* Cambridge, MA, MIT Press.

Eylon, B.-S. & Linn, M.C. (1988). 'Learning and instruction: An examination of four research perspectives in science education', *Review of Educational Research* (58), 251–301.

Fairchild, Poltrock, & Furnas, G.W. (1986). 'SemNet: Three-dimensional graphic representations of large knowledge bases', in R. Guindon (ed.), *Cognitive science and its applications to human computer interaction,* Hillsdale, NJ, Lawrence Erlbaum, 201–233.

Faletti, J. & Fisher, K.M. (1995). 'The information in relations in biology, or the unexamined relation is not worth having', in K.M. Fisher & M.R. Kibby (eds.), *Knowledge acquisition, organization, and use in biology* (NATO ASI Series F, Vol. 148), New York, Springer Verlag, 182–205.

Fegahli, A. (1991). 'A study of engineering college students' use of computer-based semantic networks in a computer programming language class', unpublished master's thesis, West Lafayette, IN, Purdue University.

Fensham, P.J., Garrard, J.E. & West, L.H.T. (1981). 'The use of cognitive mapping in teaching and learning strategies', *Research in Science Education* (11), 121–129.

Fisher K. M., Faletti, J., Thornton, R., Patterson, H., Lipson, J., & Spring, C. (1987). *Computer-based knowledge representation as a tool for students and teachers.* Technical Manuscript. Available from K. M. Fisher, CRMSE, 6475 Alvarado Road, San Diego State University, San Diego, CA 92182.

Fisher, K.M. & Faletti, J. (1989). 'Student strategies in building semantic networks in biology', paper presented at the annual meeting of the American Educational Research Association, San Francisco.

Fisher, K.M., Faletti, J., Patterson, H.A., Thornton, R., Lipson, J. & Spring, C. (1990). 'Computer-based concept mapping: SemNet software—A tool for describing knowledge networks', *Journal of College Science Teaching* (19), 347–352.

Fisher, K.M. & Faletti, J. (1993). 'Metacognition is not enough: Promoting metacognition about knowledge organization skills in biology', presented in a symposium on metacognition and conceptual change at the annual meeting of the American Educational Research Association, Atlanta, GA.

Fisher, K.M. & Gomes, S. (1996a). 'Teaching biology to prospective elementary school teachers so as to promote transition from receiver to giver of information', *Proceedings of the second international conference on the learning sciences,* Evanston, IL.

Fisher, K. M. & Gomes, S. (1996b). 'Advanced knowledge acquisition for prospective elementary biology teachers', presented in a symposium on teaching biology for meaningful learning: Constructivism in the classroom at the annual meeting of the American Association for the Advancement of Science Pacific Division, San Jose, CA.

Fisher, K.M. & Kibby, M. (eds.). (1996). *Knowledge acquisition, organization and use in biology,* Heidelberg, Springer Verlag.

Fisher, K.M. & Lipson, J.I. (1985). 'Information processing interpretation of errors in college science learning', *Instructional Science* (14), 49–74.

Fisher, K.M. (1988). 'Relations used in student-generated knowledge representations', presented in the symposium on student understanding in science: issues of cognition and curriculum at the annual meeting of the American Educational Research Association, New Orleans, LA.

Fisher, K.M. (1990). 'Semantic networking: The new kid on the block', *Journal of Research in Science Teaching* (27), 1001–1018.

Fisher, K.M. (1991). 'SemNet: A tool for personal knowledge construction,' in P. A. M Kommers, D. H. Jonassen, & J. T. Mayes (eds.), *Cognitive tools for learning,* Berlin, Springer-Verlag, 63–75.

Fisher, K.M. (1993). 'Metacognitive tools: Concept mapping', in *Proceedings of the third international seminar on misconceptions and educational strategies in science and mathematics,* Ithaca, NY, Cornell University, 18.

Fisher, K.M. (1995). 'Applying cognitive flexibility theory to the teaching of biology to prospective elementary school teachers', in a symposium on cognitive perspectives on science and mathematics education at the second international symposium on cognition and education, organized by Banaras Hindu University, Varanasi, India; Cognitive Science Centre, McGill University, Toronto; and Catholic University, Washington, DC.

Fisher, K.M. (1995). 'Models of long term memory: Evaluation of student-constructed semantic networks', paper presented at the symposium on mental models and representation in new assessments at the annual meeting of the American Educational Research Association, San Francisco.

Fisher, K.M. (1999). *Biology lessons for prospective and practicing teachers*, home page: http://www.BiologyLessons.sdsu.edu/.

Fisher, K.M. (in press a). 'SemNet® software as an assessment tool', in J.J. Mintzes, J.H. Wandersee & J.D. Novak (eds.), *Assessing science understanding: A human constructivist view*, San Diego, Academic Press.

Fisher, K.M. (in press b). in J. Minstrell & E.H. van Zee (eds.), *Teaching and learning in an inquiry-based science classroom*, Washington, DC, American Association for the Advancement of Science.

Flavell, J.H. (1977). *Cognitive development*, Englewood Cliffs, NJ, Prentice-Hall.

Flick, L.B. (1997). 'Understanding a generative learning model of instruction: A case study of elementary teacher planning', *Journal of Science Teacher Education* (7), 95–122.

Fowler, R.H., Fowler, W.A.L. & Williams, J.L. (1997). *Document explorer visualizations of WWW document and term spaces*, unpublished manuscript, Edinburgh, TX, University of Texas—Pan American.

Freeman, K. (1998, March 24). 'A shade of differences', *The New York Times* p. B16.

Fremerman, S. (1998, July–August). 'Meet Magnet, P.I.', *Natural Health* 52, 54, 56.

French, R.M. (1995). *The subtlety of sameness: A theory and computer model of analogy-making*, Cambridge, MA, MIT Press.

Friedlander, B.P. (1997, June 19). 'California first-grader gives beetle to insect collection', *The Cornell Chronicle* p. 1.

Fuller, R.G., Karplus, R. & Lawson, A. (1977). 'Can physics develop reasoning?', *Physics Today* (30), 23–28.

Furnas, G.W. (1986). 'Generalized fisheye views', *CHI Proceedings* (ACM 0-89791-180-6/86/0400-0016), 16–23.

Gagne, R.M. (1977). *The conditions of learning, third edition*, San Francisco, Holt, Rinehart, & Winston.

Gardner, E.J. (1972). *History of biology*, Minneapolis, MN, Burgess Publishing.

Gardner, M. (1968). *Logic machines, diagrams, and Boolean algebra*, New York, Dover Publications.

Garvie, L.A.J. (1994). 'A semantic net representation for the classification of minerals', *Computers & Geosciences* (21), 387–396.

Gentner, D. (1978). 'On relational meaning: The acquisition of verb meaning', *Child Development* (49), 988–998.

Gentner, D. (1981a). 'Integrating verb meanings into context', *Discourse Processes* (4), 349–375.

Gentner, D. (1981b). 'Some interesting differences between verbs and nouns', *Cognition and Brain Theory* (4), 161–178.

Gentner, D. (1982). 'Why nouns are learned before verbs: Linguistic relativity versus natural partitioning', in S. Kuczaj (ed.), *Language development: Language, cognition, and culture*, Hillsdale, NJ: Erlbaum, 301–334.

Glynn, S.M. (1989). 'The teaching with analogies model: Explaining concepts in expository texts', in K.D. Muth (ed.), *Children's comprehension of narrative and expository text: Research into practice*, Newark, DE, International Reading Association, 185–204.

Good, R.G., Trowbridge, J.E., Demastes, S.S., Wandersee, J.H., Hafner, M.S. & Cummins, C. L. (eds.) (1992). *Proceedings of the 1992 evolution education conference*, Baton Rouge, Louisiana State University.

Gordon, S.E. & Gill, R.T. (1989). *The formation and use of knowledge structures in problem solving domains* (Project Report), Moscow, ID, Psychology Department, University of Idaho.

Gordon, S.E. & Gill, R.T. (1989). *The formation and use of knowledge structures in problem solving domains* (Project Report), Moscow, ID, Psychology Department, University of Idaho.

Gordon, S.E. (1989). 'Theory and methods for knowledge acquisition', *AI Applications* 3(3), 19–30.

Gordon, S.E. (1996). 'Eliciting and representing biology knowledge with conceptual graph structures', in K.M. Fisher & M.R. Kibby (eds.), *Knowledge acquisition, organization, and use in biology* (NATO ASI Series F, Vol. 148), New York, Springer Verlag, 135–154.

Gordon, W.T. (1997). *McLuhan for beginners,* New York, Writers and Readers Publishing.

Gorodetsky, M. & Fisher, K.M. (1996). 'Generating connections and learning in biology', in K.M. Fisher & M.R. Kibby (eds.), *Knowledge acquisition, organization, and use in biology* (NATO ASI Series F, Vol. 148), New York, Springer Verlag, 135–154.

Gorodetsky, M., Fisher, K.M. & Wyman, B. (1994). 'Generating connections and learning with SemNet, a tool for constructing knowledge networks', *Journal of Science Education and Technology* (3), 137–144.

Gottfried, S.S. & Kyle, W.C.Jr. (1992). 'Textbook use and the biology education desired state', *Journal of Research in Science Teaching* (29), 35–49.

Gould, S.J. (1986). *The panda's thumb: More reflections in natural history,* New York, W.W. Norton.

Gould, S.J. (1994). 'The evolution of life on earth', *Scientific American* (271), 85–91.

Grabiner, J.V. & Miller, P.D. (1974). 'Effects of the Scopes trial', *Science* (185), 832–837.

Grabinger, R.S. & Dunlap, J.C. (1995). 'Rich environments for active learning: a definition', *ALT-J 3* (2), 5–34.

Grabinger, R.S. (1996). 'Rich environments for active learning', in D.H. Jonassen (ed.), *Handbook of research for educational communications and technology,* New York, Macmillan, 665–692.

Grabinger, S., Dunlap, J. & Duffield, J. (1997). 'Student-centered learning environments in action: Problem-based learning', *ALT-J* 5(2), 3–17.

Gravett, S.J. & Swart, E. (1997). 'Concept mapping: A tool for promoting and assessing conceptual change', *South African Journal of Higher Education* (11), 122–126.

Griffard, P.B. & Wandersee, J.H. (1998). 'Challenges to meaningful learning among African-American females at an urban science high school', paper presented at the annual meeting of the National Association for Research in Science Teaching, San Diego, CA.

Gross, P.R. (1997, December 1). 'Science without scientists', *The New York Times* p. A23.

Gunstone, R.F. (1994). 'The importance of specific science content in the enhancement of metacognition', in P. Fensham, R. Gunstone & R. White (eds.), *The content of science: A constructivist approach to its teaching and learning,* London, Falmer Press, 131–146.

Guzzetti, B.J., Snyder, T.E., Glass, G.V. & Gamas, W.S. (1993). 'Promoting conceptual change in science: A comparative meta-analysis of instructional interventions from reading education and science education', *Reading Research Quarterly* (28), 116–159.

Hackney, M.W. & Wandersee, J.H. (in press). 'Observations about the writing of Dawkins, Gould, Mayr, Thomas, & Wilson: Notes of two metaphor watchers', *Adaptation* (Journal of the New York Biology Teachers Association).

Haslam, F. & Treagust, D.F. (1987). 'Diagnosing secondary students' misconceptions of photosynthesis and respiration in plants using a two-tier multiple choice instrument', *Journal of Biological Education* (21), 203–211.

Hellmann, R.A. (1965). 'Evolution in American school biology books from the late nineteenth century until the 1930s', *The American Biology Teacher* (27), 778–780.

Helms, H. & Novak, J.D. (1983). *Proceedings of the first international seminar: Misconceptions and educational strategies in science and mathematics: Vol. I,* Ithaca, NY, Department of Education, Cornell University.

Hestenes, D. & Halloun, I. (1995). 'Interpreting the force concept inventory', *The Physics Teacher* (33), 502–506.

Hestenes, D., Wells, M. & Swackhamer, G. (1992). 'Force concept inventory', *The Physics Teacher* (30), 141–151.

Hettich, P.I. (1992). *Learning skills for college—and career,* Belmont, CA, Wadsworth.

Hewson, P.W. & Hewson, M.G. (1988). 'An appropriate conception of teaching science: A view from studies of science learning', *Science Education* (72), 597–614.

Hiebert, J., Carpenter, T.P., Fennema, E., Fuson, K.C., Wearne, D., Murray, H. & Olivier, A. (1997). *Making sense: Teaching and learning mathematics with understanding,* Portsmouth, NH, Heinemann.

Hoagland, M. & Dodson, B. (1995). *The way life works,* New York, Random House.

Hoffman, R.P. (1991). *Use of relational descriptors by experienced users of a computer-based semantic network,* unpublished master's thesis, San Diego, CA, San Diego State University.

Hofmann, J. & Welschselgartner, B. (1990a). *Guide through the HECTOR (European common market analysis) framework of reference* (Technical Report), Stuttgart, Germany, Fraunhofer-Institut für Arbeitswirtschaft und Organisation.

Hofmann, J. & Welschselgartner, B. (1990b). *Tools assessment for presenting the contents of the HECTOR (European common market analysis) framework of reference* (Technical Report), Stuttgart, Germany, Fraunhofer-Institut für Arbeitswirtschaft und Organisation.

Holden, C. (1990). Entomologists wane as insects wax. *Science, 246*, 745–755.

Holley, C.D. & Dansereau, D.F. (1984). *Spatial learning strategies,* New York, Academic Press.

Holliday, W.G., Yore, L.D. & Alvermann, D.E. (1994). 'The reading-science learning-writing connection: Breakthroughs, barriers, and promises', *Journal of Research in Science Teaching* (31), 877–893.

Horn, R. E. (1989). *Mapping hypertext:.* Lexington, MA: The Lexington Institute.

Horton, P., McConney, A., Gallo, M., Woods, A., Senn, G. & Hamelin, D. (1993). 'An investigation of the effectiveness of concept mapping as an instructional tool', *Science Education* (77), 95–111.

Horwitz, P. & Barowy, B. (1994). 'Designing and using open-ended software to promote conceptual change', *Journal of Science Education and Technology* (3), 161–185.

Howard Hughes Medical Institute (1996). *Beyond bio 101,* Chevy Chase, MD, Howard Hughes Medical Institute.

Hoz, R., Tomer, Y. & Tamir, P. (1990). 'The relations between disciplinary and pedagogical knowledge and the length of teaching experience of biology and geography teachers', *Journal of Research in Science Teaching* (27), 973–988.

Hyerle, D. (1996). *Visual tools for constructing knowledge,* Alexandria, VA, Association for Supervision and Curriculum Development.

Jackson, D.F., Doster, E.C., Meadows, L. & Wood, T. (1995). 'Hearts and minds in the science classroom: The education of a confirmed evolutionist', *Journal of Research in Science Teaching* (32), 585–611.

Jacobson, M.J. & Spiro, R.J. (1995). 'Hypertext learning environments, cognitive flexibility, and the transfer of complex knowledge', *Journal of Educational Computing Research* (12), 301–333.

Jay, M., Alldredge, S. & Peters, F. (1990). 'SemNet preliminary classroom study: 7th grade biology (ecology)', presented at the annual meeting of the American Educational Research Association, Boston.

Jeannerod, M. (1985). *The brain machine: The development of neurophysiological thought,* Cambridge, MA, Harvard University Press.

Johnson, D.W. & Johnson, R. (1975/1991). *Learning together and alone: Cooperative, competitive, and individualistic learning,* Englewood Cliffs, NJ, Prentice-Hall.

Johnson, D.W. (1974). 'Communication and the inducement of cooperative behavior in conflicts: A critical review', *Speech Monographs* (41), 64–78.

Johnson, D.W., & Johnson, R. (1983). 'The socialization and achievement crisis: Are cooperative learning experiences the solution?', in L. Bickman (ed.), *Applied social psychology annual 4,* Beverly Hills, CA, Sage, 119–164.

Johnson, D.W., Johnson, R. & Smith, K. (1991). *Active learning: cooperation in the college classroom,* Edina, MN, Interaction Book Company.

Jonassen, D.H., Beissner, K. & Yacci, M. (1993). *Structural knowledge: Techniques for representing, conveying and acquiring structural knowledge,* Hillsdale, NJ, Lawrence Erlbaum.

Jonassen, D.W. & Wang, S. (1993). 'Acquiring structural knowledge from semantically structured hypertext', *Journal of Computer-Based Instruction* 20(1), 19–28.

Kelly, G.A. (1955). *The psychology of personal constructs: vol. 1, A theory of personality,* New York, W.W. Norton.

Kliebard, H.M. (1965). 'Structure of the disciplines as an educational slogan', *Teachers College Record* (66), 598–603.

Komorowski, H.J. & Greenes, R.A. (1987). 'The use of fisheye views for displaying semantic relationships in a medical taxonomy', in W.W. Stead (ed.), *Proceedings of the 11th annual symposium on computer applications in medical care,* New York, Institute of Electrical and Electronic Engineers, 113–116.

Kourik, R. (1997, November 30).'In the garden, science isn't always exact', *The New York Times* p. 37.

Kozhenikov, M., Hegarty, M. & Mayer, R. (1999). 'The role of imagery in problem solving in physics', presented at the annual meeting of the American Educational Research Association, Montreal, Canada.

Kuhn, D. (1999). A developmental model of critical thinking. *Educational Researcher* 28(2), 16–26, 46.

Kuhn, T.S. (1970). *The structure of scientific revolutions, second edition*, Chicago, The University of Chicago Press.

Lakoff, G. & Johnson, M. (1981). *Metaphors we live by*. Chicago: The University of Chicago Press.

Lakoff, G. (1987). *Women, fire, and dangerous things: What categories reveal about the mind*, Chicago, University of Chicago Press.

Lambiotte, J.G., Dansereau, D.F., Cross, D.R. & Reynolds, S.B. (1989). 'Multirelational semantic maps', *Educational Psychology Review* (1), 331–367.

Langer, E.J. (1989). *Mindfulness*, Menlo Park, CA, Addison-Wesley.

Langer, E.J. (1997). *The power of mindful learning*, Menlo Park, CA, Addison-Wesley.

Langer, S. (1988). *Mind: An essay on human feelings*, Baltimore, John Hopkins University Press.

Lanzing, J.W.A. (1998). 'Concept mapping: Tools for echoing the mind's eye', *Journal of Visual Literacy* 18(1), 1–14.

Larkin, J. & Simon, H. A. (1987). 'Why a diagram is (sometimes) worth ten thousand words', *Cognitive Science* (11), 65–99.

Lave, J., Murtaugh, M. & de la Rocha, O. (1984). 'The dialectic of arithmetic in grocery shopping', in B. Rogoff & J. Lave (eds.), *Everyday cognition: Its development in social contexts*, Cambridge, Harvard University Press.

Lavoie, D.R. (1998). 'The cognitive-processing nature of concept mapping in biology', paper presented at the annual meeting of the National Association for Research in Science Teaching, San Diego, CA.

Lehman, F. (1992). Semantic networks, In F. Lehman & E.Y. Rodin (eds.), *Semantic networks in artificial intelligence*, New York, Pergamon Press.

Lehman, F. (ed.) & Rodin, E.Y. (general ed.). (1992). *Semantic networks in artificial intelligence*, New York, Pergamon Press.

Levy, G.B. (1987). '...What is out front', *American Laboratory* 19(4), 10.

Lieberman, P. (1997). 'Peak capacity', *The Sciences* 37(6), 22–27.

Linder, C.J. (1993). 'A challenge to conceptual change', *Science Education* (77), 292–300.

Liu, M. (1993). 'The effects of hypermedia instruction on second language learning through a semantic-network-based approach', paper presented at the annual conference of Eastern Educational Research Association, Clearwater, FL.

Lloyd, C.V. (1990). 'The elaboration of concepts in three biology textbooks: Facilitating student learning', *Journal of Research in Science Education* (27), 1019–1032.

Locke, J. (1690/1996). *Essay concerning human understanding* (abridged and edited, with an introduction and notes by K.P. Winkler), Indianapolis, IN, Hackett Publishing.

Locke, J. (1706/1891). *The conduct of the understanding* (with introduction and notes by J.A. St. John), New York, Alden.

Logan, J. (1999). *World music in contemporary life (Music 345)*, College of Extended Studies, San Diego State University, Knowledge webs home page available at http://trumpet.sdsu.edu/M345/Knowledge_Webs/knowledge_webs.html

Longo, P.J. (1999). 'Visual thinking network—A new generation of dimensional representation strategies for learning science', paper presented at the annual meeting of the American Educational Research Association, Montreal, Canada.

Lovitt, Z. & Burk, J. (1988). 'Webbing: A bridge between teaching and learning', *Educational Horizons* (66), 119–121.

Luoma-Overstreet, K. (1990). 'SemNet Journal: A documentation of progress over the duration of the final assignment', submitted for Cognitive Science 700 to Dr. B. Allen, San Diego State University, San Diego, CA.

Macnamara, J. (1984). *Names for things: A study of human learning*, Cambridge, MA, MIT Press.

Magner, L.N. (1979). *A history of the life sciences*, New York, Marcel Dekker.

Markman, E.M. (1989). *Categorization and naming in children: Problems of induction*, Cambridge, MA, MIT Press.

Marra, R.M. (1997). 'The effects of generating semantic networks on knowledge synthesis as measured by expert system creation' (Doctoral dissertation, University of Colorado at Denver, 1996), *Dissertation Abstracts International* (58), 138A.

Mason, L. (1994). 'Cognitive and metacognitive aspects in conceptual change by analogy', *Instructional Science* (22), 157–187.

Mastrilli, T.M. (1997). 'Instructional analogies used by biology teachers: Implications for practice and teacher preparation', *Journal of Science Teacher Education* (8), 187–204.

Mayer, R.E. & Sims, V.K. (1994). 'For whom is a picture worth a thousand words? Extensions of the dual-coding theory of multimedia learning', *Journal of Educational Psychology* (46), 389–401.

Mayr, E. (1982). *The growth of biological thought: Diversity, evolution and inheritance*, Cambridge, MA, Harvard University Press.

McAleese, R. (1985). 'Some problems of knowledge representation in an authoring environment: Exteriorization, anomalous state meta-cognition and self-confrontation', *Programmed Learning and Educational Technology* (22), 299–306.

McAleese, R., Grabinger, S. & Fisher, K. (1999). 'The knowledge arena: A learning environment that underpins concept mapping', presented at the annual meeting of the American Educational Research Association, Montreal, Canada.

McComas, W.F. (1997). '15 myths of science: Lessons of misconceptions and misunderstandings from a science editor', *Skeptic* 5(2), 88–95.

McNeill, D. & Freiberger, P. (1993). *Fuzzy Logic: The revolutionary computer technology that is changing our world*. New York: Simon & Schuster.

Meighan, M., Wan, E. & Starratt, S. (1996). 'BioA2Z: A dynamic glossary', in the CD-ROM-based *Biology survival kit*, Philadelphia, PA, Saunders College Publishing.

Miller, G. A., Beckwith, R., Fellbaum, C., Gross, D. & Miller, K. (1990) 'Introduction to WordNet: An on-line lexical database', in Princeton University (ed.), *Cognitive science laboratory: Five papers on WordNet*, Princeton, NJ, Princeton University, 1–10.

Miller, G.A. (1956). 'The magical number seven, plus or minus two: Some limits on our capacity for information processing', *Psychological Review* (63), 81–97.

Miller, J. G. (1978). *Living Systems*. New York: McGraw-Hill.

Miller, J.G. (1973). 'Living systems', *The Quarterly Review of Biology* (48), 63–276.

Mintzes, J.J., Wandersee, J.H. & Novak, J.D. (1997). 'Meaningful learning in science: The human constructivist perspective', In G. Phye (ed.), *Handbook of academic learning: Construction of knowledge*, San Diego, CA, Academic Press, 405–447.

Moody, D.E. (1993). 'Insight as the basis for a functional typology of misconceptions', paper presented at the third international conference on misconceptions and educational strategies in science and mathematics, Ithaca, NY, Cornell University.

Moody, D.E. (1996). 'Evolution and the textbook structure of biology', *Science Education* (80), 395–418.

Moore, J.A. (1993). *Science as a way of knowing: The foundations of modern biology*, Cambridge, MA, Harvard University Press.

Morine-Dershimer, G. (1993). 'Tracing conceptual change in preservice teachers', *Teaching & Teacher Education* (9), 15–26.

Murray, B. (1997, August). 'Fluid, flexible thinking boosts our learning ability', in *APA Monitor*, Washington, DC, American Psychological Association.

Muschamp, H.T. (1997, December 1). 'A mountaintop temple where art's future worships its past', *The New York Times* pp. A1, A16.

Nachmias, R., Stavy, R. & Avrams, R. (1990). 'A microcomputer-based diagnostic system for identifying students' conceptions of heat and temperature', *International Journal of Science Education* (12), 123–132.

National Research Council (1990). *Fulfilling the promise: Biology education in the nation's schools*, Washington, DC, National Academy Press.

National Research Council (1996). *National science education standards*, Washington, DC, National Academy Press.

National Research Council Commission on Life Science, Board on Biology, Committee on High School Biology Education (1990). *Fulfilling the promise: Biology education in the nation's schools*, Washington, DC, National Academy Press.

National Science Board Commission on Precollege Education in Mathematics, Science and Technology (1982). *Today's problems, tomorrow's crises*, Washington, DC, National Science Foundation.

National Science Board on Precollege Education in Mathematics, Science and Technology (1983). *Educating Americans for the 21st century: A plan of action for improving mathematics, science, and technology education for all American elementary and secondary students so that their achievement is the best in the world by 1995* (NSF Publication No. CPCE-NSF-03), Washington, DC, National Science Foundation.

National Science Board Task Committee (1986). *Undergraduate science and engineering education,* Washington, DC, National Science Board.

National Science Foundation, Directorate for Education and Human Resources, Division of Elementary, Secondary and Informal Education (undated, ~1996/1997). *Foundations: The challenge and promise of K–8 science education reform,* Washington, DC, National Science Foundation.

Nelkin, D. (1977). *Science textbook controversies and the politics of equal time,* Cambridge, MA, The MIT Press.

Nichols, M.S. (1994). 'A cross-age study of students' knowledge of insect metamorphosis: Insights into their understanding of evolution' (Doctoral dissertation, Louisiana State University, 1993), *Dissertation Abstracts International* (55), 524A.

Nobles, C.S. (1993). 'Concept circle diagrams: A metacognitive learning strategy to enhance meaningful learning in the elementary science classroom' (Doctoral dissertation, Louisiana State University, 1994), *Dissertation Abstracts International* (54), 3312A.

Norman, D.A. (1981). 'Categorization of action slips', *Psychological Review* (88), 1–15.

Norman, D.A. (1993). *Things that make us smart,* Menlo Park, CA, Addison-Wesley.

Norman, D.A., Rumelhart, D.E. & the LNR Research Group (1975). *Explorations in cognition,* San Francisco, Freeman.

Novak, J. & Gowin, D.B. (1984). *Learning how to learn,* Cambridge, UK, Cambridge University Press.

Novak, J.D. & Musonda, D. (1991). Tracing conceptual change in preservice teachers. A twelve-year longitudinal study of science concept learning. *American Educational Research Journal, 28,* 117–153.

Novak, J.D. & Wandersee, J.H. (eds.). (1990). 'Special issue: Perspectives on concept mapping', *Journal of Research in Science Teaching* (27), 919–1079.

Novak, J.D. (1964). 'The importance of conceptual schemes for science teaching', *The Science Teacher* 31(6), 10.

Novak, J.D. (1987). *Proceedings of the second international seminar: Misconceptions and educational strategies in science and mathematics: Vol. II,* Ithaca, NY, Cornell University.

Novak, J.D. (1991a). 'Concept maps and Vee diagrams: Two metacognitive tools to facilitate meaningful learning', *Instructional Science* (19), 1–25.

Novak, J.D. (1991b). 'Clarify with concept maps: A tool for students and teachers alike', *The Science Teacher* (58), 45–49.

Novak, J.D. (1993). *Proceedings of the third international seminar: Misconceptions and educational strategies in science and mathematics: Vol. III,* Ithaca, NY, Cornell University.

Novak, J.D. (1998). *Learning, creating, and using knowledge: Concept maps™ as facilitative tools in schools and corporations,* Mahwah, NJ, Lawrence Erlbaum.

Odom, A.L. & Barrow, L.H. (1995). 'The development and application of a two-tiered diagnostic test measuring college biology students' understanding of diffusion and osmosis following a course of instruction', *Journal of Research in Science Teaching* (32), 45–61.

Okebukola, P. & Jegede, O. (1998). 'Nigerian teachers' perception of concept mapping and Vee-diagramming as metalearning tools in science', paper presented at the annual meeting of the National Association for Research in Science Teaching, San Diego, CA.

Okebukola, P.A. (1990). 'Attaining meaningful learning of concepts in genetics and ecology; an examination of the potency of the concept-mapping technique', *Journal of Research in Science Teaching* (27), 493–504.

Ortony, A. (ed.). (1979). *Metaphor and thought,* New York, Cambridge University Press.

Osborne, R. & Gilbert, J. (1980). 'A method for investigating concept understanding in science', *European Journal of Science Education* (2), 311–321.

Otto, J.H., & Towle, A. (1985). *Modern Biology,* New York: Holt, Rinehart, & Winston.

Pagels, H.R. (1988). *The dreams of reason: The computer and the rise of the sciences of complexity,* New York, Bantam Books.

Paivio, A. (1991). 'Dual coding theory: Retrospect and prospect', *Canadian Journal of Psychology* (45), 255–287.

Papert, S. (1980). *Mindstorms: Children, computers and powerful ideas,* New York, Basic.

Pask, G. (1975). *Conversation, cognition and learning: A cybernetic theory and methodology,* New York, Elsevier.

Pask. G. (1976a). Conversational techniques in the study and practice of education. *Journal of Educational Psychology, 46*, 12–25.

Pask. G. (1976b). Styles and strategies of learning. *Journal of Educational Psychology, 46*, 128–248.

Pask, G. (1977). Learning styles, educational strategies, and representations of knowledge: Methods and applications. *SSRC Research Programme HR2708/1 Progress Report 3, Volume 1 & 2*. Richmond, Surrey, GB: System Research Ltd., 37 Sheen Road, Richmond, Surrey.

Pea, R.D. & Gomes, L.M. (1992). 'Distributed multimedia learning environments: Why and how?', *Interactive Learning Environments* (2), 73–109.

Pea, R.D. (1985). 'Beyond amplification: Using the computer to reorganize mental functioning', *Educational Psychologist* (20), 167–182.

Peirce, C.S. (1931–1958). *Collected papers of Charles Sanders Peirce, Vol. 4* (edited by C. Hartshorne, P. Weiss & A. Burks), Cambridge, MA, Belknap Press (Harvard University Press).

Peirce, C.S. (1976). *The new elements of mathematics, Vol III/1* (edited by C. Eisele), Atlantic City, NJ, Mouton/Humanities Press/The Hague.

Perkins, D.N. & Salomon, G. (1989). 'Are cognitive skills context-bound?', *Educational Researcher* 18(1), 16–25.

Perkins, D.N. (1993). 'Person-plus: A distributed view of thinking and Learning', in G. Salomon (ed.), *Distributed cognitions: Psychological and educational considerations*, New York, Cambridge University Press.

Pfundt, H. & Duit, R. (1994). *Bibliography: Students' alternative frameworks and science education, fourth edition*, Kiel, Germany, Institut für die Pedagogic der Naturwissenschaften (Institute for Science Education), Universität Kiel.

Pines, A., Novak, J.D., Posner, G. & VanKirk, J. (1978). *The clinical interview: A method for evaluating cognitive structure* (Research Report #6), Ithaca, NY, Department of Education, Cornell University.

Pinker, S. (1994). *The language instinct: How the mind creates language*, New York, Harper: HarperPerenial.

Pope, M. & Gilbert, J. (1983). 'Conceptual understanding and science learning: An interpretation of research within a sources-of-knowledge framework', *Science Education* (70), 583–604.

Pope, M. (1982). 'Personal construction of formal knowledge', *Interchange on Educational Policy* (13), 4.

Posner, G.J., Strike, K.A., Hewson, P.W. & Gerzog, W.A. (1982). 'Accommodation of a scientific conception: Toward a theory of conceptual change', *Science Education* (66), 211–227.

Quillian, M.R. (1967). 'Word concepts: A theory and simulation of some basic semantic capabilities', *Behavioral Sciences* (12), 410–430.

Quillian, M.R. (1968). 'Semantic memory', in M. Minsky (ed.), *Semantic information processing*, Cambridge, MA, MIT Press, 216–270.

Quillian, M.R. (1969). 'The teachable language comprehender', *Communications of the Association for Computing Machinery* (12), 459–475.

Reader, W. & Hammond, N. (1998). *Computer-based tools to support learning from hypertext: Concept mapping tools and beyond*, available from the internet: http://www.ioc.ac.uk/tescwwr/CAL.html

Reif, F. & Heller, J.I. (1981). *Knowledge structure and problem solving in physics* (Educational Science Paper No. 12), Berkeley, CA, University of California.

Reif, F. & Heller, J.I. (1982). *Making scientific concepts and principles effectively usable: Requisite knowledge and teaching implications* (Educational Science Paper No. ES-13), Berkeley, CA, University of California.

Reif, F. & Larkin, J. (1991). 'Cognition in scientific and everyday domains: Comparisons and learning implications', *Journal of Research in Science Teaching* (28), 733–760.

Reif, F. (1983). 'Understanding and teaching problem solving in physics', in *Research on physics education: Proceedings of the first international workshop*, Paris, France, La Londe les Maures.

Reiger, C. (1976). 'An organization of knowledge for problem-solving and language. comprehension', *Artificial Intelligence* (7), 89–127.

Reisbeck, C.K. (1975). *Conceptual information processing*, New York, American L. Elsevier.

Resnick, L.B. (1987a). 'Learning in school and out', *Educational Researcher* 16(9), 13–20.

Resnick, L.B. (1987b). *Education and learning to think*, Washington, DC, National Academy Press.

Reynolds, L. & Simmonds, D. (1981). *Presentation of data in science*, The Hague, The Netherlands, Martinus Nijhoff Publishers.

Richens, R.H. (1956). 'Preprogramming for mechanical translation', *Mechanical Translation* 3(1), a discontinued journal available at the Philip Mills Arnold Semiology Collection, Rare Book Department, Olin Library, Washington University, St. Louis, MO 63130.

Rico, G.L. (1983). *Writing the natural way*, Los Angeles, J. P. Tarcher.

Robinson, A. (1993). *What smart students know*, New York, Crown Publishers.

Rogoff, B. & Lave, J. (eds.). (1984). *Everyday cognition: Its development in social contexts*, Cambridge, Harvard University Press.

Rosch, E. & Lloyd, B.B. (eds.). (1978). *Cognition and categorization*, Hillsdale, NJ, Lawrence Erlbaum.

Rosch, E. & Mervis, C. (1975). 'Family resemblances: Studies in the internal structures of categories', *Cognitive Psychology* (7), 573–605.

Rosch, E. (1973). 'On the internal structure of perceptual and semantic categories' in T.E. Moore (ed.), *Cognitive development and the acquisition of language*, New York, Academic Press.

Rosch, E. (1975). 'Cognitive representations of semantic categories', *Journal of Experimental Psychology: General* (104), 192–223.

Rosch, E. (1977). 'Human categorization', in N. Warren (ed.), *Advances in cross-cultural psychology, 1*, London, Academic Press.

Rosch, E., Mervis, C., Gray, W., Johnson, D. & Boyes-Braem, P. (1976). 'Basic objects in natural categories', *Cognitive Psychology* (8), 382–439.

Rosenberg, A. (1985). *The structure of biological science*, New York, Cambridge University Press.

Rosenthal, D.B. & Bybee, R.W. (1986). 'Emergence of the biology curriculum: A science of life or a science of living?', in T.S. Popkewitz (ed.), *The formation of school subjects: The struggle for creating an American institution*, New York, Falmer Press, 123–144.

Rosenthal, J.W. (1996). *Teaching science to language minority students*, Bristol, PA, Multilingual Matters Ltd.

Roth, K. & Anderson, C. (1988). 'Promoting conceptual change learning from science textbooks', in P. Ramsden (ed.), *Improving learning: New perspectives*, New York, Nichols Publishing, 109–141.

Ruiz-Primo, M.A. & Shavelson, R.J. (1996). 'Problems and issues in the use of concept maps in science assessment', *Journal of Research in Science Teaching* (33), 569–600.

Rumelhart, D.E. & Norman, D.A. (1978). 'Accretion, tuning, and restructuring: Three modes of learning', in J.W. Cotton & R. Klayzky (eds.), *Semantic factors in cognition*, Hillsdale, NJ, Erlbaum, 37–53.

Rumelhart, D.E. & Ortony, A. (1977). 'The representation of knowledge in memory', in R.C. Anderson, R.J. Spiro & W.E. Montague (eds.), *Schooling and the acquisition of knowledge*, Hillsdale, NJ, Lawrence Erlbaum.

Rumelhart, D.E. (1980). 'Schemata: The building blocks of cognition', in R.J. Spiro, B.C. Bruce & W.F. Brewer (eds.), *Theoretical issues in reading comprehension: Perspectives from cognitive psychology, linguistics, artificial intelligence, and education*, Hillsdale, NJ, Lawrence Erlbaum.

Rumelhart, D.E., McClelland, J.L. & The PDP Research Group (1987). *Parallel distributed processing: Explorations in the microstructure of cognition. Volume 1: Foundations*, Cambridge, MA, MIT Press.

Sagan, C. (1979). *Broca's brain*, New York, Random House.

Sakurai, J. (1997, November 29). 'Sorry, you're not my type', *The San Diego Union-Tribune* p. A-27.

Salomon, G. & Globerson, T. (1987). 'Skill may not be enough: The role of mindfulness in learning and transfer', *International Journal of Education Research* (11), 623–638.

Schmidt, W.H., McKnight, C.C. & Raizen, S.A. (1996). *A splintered vision: An investigation of U.S. science and mathematics education*, Boston, Kluwer Academic Publishers.

Schnepps, M. (Producer). (1994). *Why are some ideas so difficult?* (video program), Cambridge, MA, Science Media Group, Harvard/Smithsonian Center for Astrophysics.

Schnepps, M. (Producer). (1997a). *Minds of our own: Can we believe our eyes?* (video program for public television), Cambridge, MA, Science Media Group, Harvard/Smithsonian Center for Astrophysics.

Schnepps, M. (Producer). (1997b). *Minds of Our Own: Lessons from thin air* (video program for public television), Cambridge, MA, Science Media Group, Harvard/Smithsonian Center for Astrophysics.

Schvaneveldt, R.W. (1990a). 'Proximities, networks, and schemata', in R. Schvaneveldt (ed.), *Pathfinder associative networks: Studies in knowledge organization*, Norwood, NJ, Ablex, 135–148.

Schvaneveldt, R.W. (1990b). *Pathfinder associative networks: Studies in knowledge organization*, Norwood, NJ, Ablex.

Shea, C. (1996, September 27). 'Researchers try to understand why people are doing better on IQ tests', *The Chronicle of Higher Education* p. A18.

Shepard, R.N. & Cooper, L.A. (1982). *Mental images and their transformations,* Cambridge, MA, MIT Press.

Shymansky, J.A., Yore, L.D. & Good, R. (1991). 'Elementary school teachers' beliefs about and perceptions of elementary school science, science reading, science textbooks, and supportive instructional factors', *Journal of Research in Science Teaching* (28), 437–454.

Simon, H.A. (1962). 'The architecture of complexity', *Proceedings of the American Philosophical Society* (106), 469–482.

Simon, H.A. (1974). 'How big is a chunk?', *Science* (183), 482–488.

Skoog, G. (1979). 'Topic of evolution in secondary school biology textbooks: 1900–1977', *Science Education* (63), 621–640.

Skoog, G. (1984). 'The coverage of evolution in high school biology textbooks published in the 1980s', *Science Education* (68), 117–128.

Slavin, R. (1983). *Cooperative learning,* New York, Longman.

Snow, R.E., Corno, L. & Jackson, D. (1996). 'Individual differences in affective and conative functions', in D.C. Berliner & R.C. Calfee (eds.), *Handbook of educational psychology,* New York, Macmillan, 243–310.

Songer, C.J. & Mintzes, J.J. (1994). 'Understanding cellular respiration: An analysis of conceptual change in college biology', *Journal of Research in Science Teaching* (31), 621–637.

Sowa, J. (ed.). (1983). *Conceptual structures: Information processing in mind and machine,* Menlo Park, CA, Addison-Wesley.

Sowa, J. (ed.). (1990). *Principles of semantic networks: Explorations in the representation of knowledge,* San Francisco, Morgan Kaufmann

Sowa, J.F. (ed.). (1999). *Knowledge representation: Logical, philosophical, and computational foundations,* Pacific Grove, CA, PWS Publishing.

Spence, M. (1992). *Miller analogies test preparation guide,* Lincoln, NE, Cliffs Notes.

Spiegel, G.F. & Barufaldi, J.P. (1994). 'The effects of a combination of text structure awareness and graphic postorganizers on recall and retention of science knowledge', *Journal of Research in Science Teaching* (31), 913–932.

Spiro, R.J. & Nix, D. (eds.). (1990). *Cognition, education, and multimedia: Exploring ideas in high technology,* Hillsdale, NJ, Lawrence Erlbaum.

Spiro, R.J., Coulson, R.L., Feltovich, P.J. & Anderson, D.K. (1988). 'Cognitive flexibility theory: Advanced knowledge acquisition in ill-structured domains', in *Tenth annual conference of the Cognitive Science Society,* Hillsdale, NJ, Erlbaum, 375–383.

Spiro, R.J., Feltovich, P.L., Jacobson, M.J. & Coulson, R.L. (1991) 'Knowledge representation, content specification, and the skill in situation-specific knowledge assembly. Some constructivist issues as they relate to cognitive flexibility theory', *Educational Technology* 31(9), 22–25.

Starr, M. & Krajcik, J. (1990). 'Concept maps as a heuristic for science curriculum development: Toward improvement in process and product', *Journal of Research in Science Teaching* (27), 987–1000.

Stepans, J. (1985). 'Biology in elementary school: Children's conceptions of life', *The American Biology Teachers* (47), 222–225.

Stevens, S.S. (1975). *Psychophysics,* New York, John Wiley.

Stevenson, H.W. & Stigler, J. (1992). *The learning gap: Why our schools are failing and what we can learn from Japanese and Chinese education,* New York, Summit Books.

Stewart, J. Van Kirk, J. & Rowell, R. (1979). 'Concept maps: A tool for use in biology teaching', *The American Biology Teacher* (41), 171–175.

Stofflett, R.T. & Stoddart, T. (1994). 'The ability to understand and use conceptual change pedagogy as a function of prior content learning experience', *Journal of Research in Science Teaching* (31), 31–51.

Strike, K.A. & Posner, G.J. (1985). 'A conceptual change view of learning and understanding', in L.H.T. West & A.L. Pines (eds.), *Cognitive structure and conceptual change,* New York, Academic Press.

Tamir, P., Gal-Choppin, K. & Nussinovitz, R. (1981). 'How do intermediate and junior high school students conceptualize living and non-living?', *Journal of Research in Science Teaching* (18), 241–248.

Thomson, E.A. (1998, March 18). 'Team building humanoid robot', *MIT Tech Talk* pp. 5–6.

Thorley, N.R. & Stofflett, R.T. (1996). 'Representation of the conceptual change model in science teacher education', *Science Education* (80), 317–339.

Thro, M.P. (1976). 'Relationship between cognitive structure and content structure of selected physics concepts' (Doctoral dissertation, Washington University, 1976), *Dissertation Abstracts International* (37), 5018A.

Thro, M.P. (1978). 'Relationships between associative and content structure in physics', *Journal of Educational Psychology* (70), 971–978.

Tillema, H.H. & Knol, W.E. (1997). 'Promoting student teacher learning through conceptual change or direct instruction', *Teaching and Teacher Education* (13), 579–595.

TIMSS, the Third International Mathematics and Science Study. Site index available at http://wwwcsteep.bc.edu/timss1/SiteIndex.html

Tobias, S. & Everson, H.T. (in press). 'Assessing metacognitive knowledge monitoring', in G. Schraw (ed.), *Issues in the measurement of metacognition*, Lincoln NE, Buros Institute of Mental Measurements and Mahwah, NJ, Erlbaum.

Towle, A. (1993). *Modern Biology*, New York: Holt, Rinehart, & Winston.

Triangle Coalition for Science and Technology Education (1997). 'Scientists protest exclusion from standards writing', *Triangle Coalition Electronic Bulletin* 3(41), 5.

Troost, C.J. (1968). 'Evolution in biological education prior to 1960', *Science Education* (52), 300–301.

Trowbridge, J.E. & Wandersee, J.H. (1994). 'Using concept mapping in a college course on evolution: Identifying critical junctures in learning', *Journal of Research in Science Teaching* (31), 459–475.

Trowbridge, J.E. & Wandersee, J.H. (1996). 'How do graphics presented during college biology lessons affect students' learning? A concept map analysis', *Journal of College Science Teaching* (26), 54–57.

Trowbridge, J.E. & Wandersee, J.H. (1998). 'Theory-driven graphic organizers', in J.J. Mintzes, J.H. Wandersee & J.D. Novak (eds.), *Teaching science for understanding: A human constructivist view*, San Diego, CA, Academic Press, 95–131.

Tufte, E.R. (1990). *Envisioning information*, Cheshire, CT, Graphics Press.

Tyson-Bernstein, H. (1990). *A conspiracy of good intentions: America's textbook fiasco*, Washington, DC, Council for Basic Education.

von Glasersfeld, E. (1984). 'An introduction to radical constructivism', in P. Watzlawick (ed.), *The invented reality*, New York, Norton, 17–40.

von Glasersfeld, E. (1987). 'Learning as a constructive activity', in C. Janvier (ed.), *Problems of representation in the teaching and learning of mathematics*, Hillsdale, NJ, Lawrence Erlbaum, 215–227.

von Glasersfeld, E. (1993). 'Questions and answers about radical constructivism', in K. Tobin (ed.), *The practice of constructivism in science education*, Washington, DC, AAAS Press, 23–38.

Vosniadou, S. & Saljo, R. (eds.) (1994). 'Conceptual change in the physical sciences', *Learning and Instruction* (4), 1–121.

Vygotsky, L.S. (1978). *Mind in society: The development of higher psychological processes*, (edited by M. Cole, V. John-Steiner, S. Scriber & E. Souberman), Cambridge, MA, Harvard University Press.

Wade, R.C. (1994). 'Conceptual change in elementary social studies: A case study of fourth graders' understanding of human rights', *Theory and Research in Social Education* (22), 74–95.

Waldholz, M. (1997). *Curing cancer: Solving one of the greatest medical mysteries of our time*, New York, Simon & Schuster.

Walker, D.F. (1981). 'Learning science from textbooks: Toward a balanced assessment of textbooks in science education', in J.T. Robinson (ed.), *Research in science education: New questions, new directions*, Columbus, OH, ERIC Clearinghouse for Science, Mathematics, and Environmental Education, 5–20.

Wallace, J. & Mintzes, J. (1990). 'The concept map as a research tool: Exploring conceptual change in biology', *Journal of Research in Science Teaching* (27), 1033–1052.

Wandersee, J.H. & Abrams, E. (1993). 'Construction of a concept map in a clinical interview setting', *AERA Subject Matter Knowledge and Conceptual Change Newsletter* (20), 4–5.

Wandersee, J.H. (1983). 'Suppose a world without science educators', *Journal of Research in Science Teaching* (20), 711–712.

Wandersee, J.H. (1984). 'Why can't they understand how plants make food? Students' misconceptions about photosynthesis', *Adaptation* 6(1), 3, 13, 17.

Wandersee, J.H. (1987). 'Drawing concept circles: A new way to teach and test students', *Science Activities* 24(4), 1, 9–20.

Wandersee, J.H. (1988). 'Ways students read texts', *Journal of Research in Science Teaching* (25), 69–84.

Wandersee, J.H. (1990). 'Concept mapping and the cartography of cognition', *Journal of Research in Science Teaching* (27), 1069–1075.

Wandersee, J.H. (1992a). 'A standard format for concept maps', invited NARST section paper presented at the national convention of the National Science Teachers Association, Boston.

Wandersee, J.H. (1992b). 'The historicality of cognition: Implications for science education research', *Journal of Research in Science Teaching* (29), 423–434.

Wandersee, J.H. (1996). 'The graphic representation of biological knowledge: Integrating words and images', in K.M. Fisher & M.R. Kibby (eds.), *Knowledge acquisition, organization, and use in biology* (NATO ASI Series F, Vol. 148), New York, Springer Verlag, 25–35.

Wandersee, J.H., Mintzes, J.J. & Novak, J.D. (1994). 'Research on alternative conceptions in science', in D.L. Gabel (ed.), *Handbook of research on science teaching and learning*, New York, Simon & Schuster MacMillan, 177–210.

Watson, J. (1991),*The Double Helix: A Personal Account of the Discovery and Structure of DNA*, New York, Penguin Group.

Weinberg, S.L. (1978). 'Two views on the textbook watchers', *The American Biology Teacher* (40), 541–545.

Weiner, J. (1995), *The Beak of the Finch: A story of evolution in our time*, New York, Vintage Books.

Weiss, I.R. (1978). *Report of the 1977 national survey of science, mathematics, and social studies education: Center for educational research and evaluation*, Washington, DC, U.S. Government Printing Office.

Weiss, I.R. (1987). *Report of the 1985–86 national survey of science and mathematics education*, Research Triangle Park, NC, Center for Educational Research and Evaluation, Research Triangle Institute.

Weitzman, E.A. & Miles, M.B. (1995). *A software sourcebook: Computer programs for qualitative data analysis*, Thousand Oaks, CA, Sage.

West, L.H.T. & Pines, A.L. (1985). *Cognitive structure and conceptual change*, New York, Academic Press.

West, L.H.T., Fensham, P.J. & Garrard, J.E. (1985). 'Describing the cognitive structures of learners following instruction in chemistry', in L.H.T. West & A.L. Pines (eds.), *Cognitive structure and conceptual change*, New York, Academic Press.

White R. & Gunstone, R. (1989). 'Metalearning and conceptual change', *International Journal of Science Education* (11), 577–586.

White, R. & Gunstone, R. (1992). *Probing understanding*, London, Falmer Press.

Wilford, J.N. (1998). 'Revolutions in mapping', *National Geographic* 193(2), 6–39.

Willerman, M. & MacHarg, R.A. (1991). 'The concept map as an advance organizer', *Journal of Research in Science Teaching* (28), 705–711.

Wilson, E.O. (1998). *Consilience: The unity of knowledge*, New York, Alfred A. Knopf.

Windschitl, M. & Andre, T. (1998). 'Using computer simulations to enhance conceptual change: The roles of constructivist instruction and student epistemological beliefs', *Journal of Research in Science Teaching* (35), 145–160.

Windschitl, M. (1997). 'Student epistemological beliefs and conceptual change activities: How do pair members affect each other?', *Journal of Science Education and Technology* (6), 37–47.

Winner, E. (1988). *The point of words: Children's understanding of metaphor and irony*, Cambridge, MA, Harvard University Press.

Wittrock, M.C. (1974a). 'A generative model of mathematics learning', *Journal for Research in Mathematics Education* (5), 181–196.

Wittrock, M.C. (1974b). 'Learning as a generative process', *Educational Psychologist* (11), 87–95.

Woodburn, J.H. & Obourn, E.S. (1965). *Teaching the pursuit of science*, New York, MacMillan.

Yore, L.D. (1991). 'Secondary science teachers' attitudes toward and beliefs about science reading and science textbooks', *Journal of Research in Science Teaching* (28), 55–72.

Zadeh, L.A. (1963). 'A computational approach to fuzzy quantifiers in natural languages', *Computers & Mathematics With Applications* 9(1), 149–184.

Zadeh, L.A. (1973). 'Outline of a new approach to the analysis of complex systems and decision processes', *IEEE Transactions on Systems, Man, & Cybernetics* (3), 28–44.

Zadeh, L.A. (1976). 'A fuzzy-algorithmic approach to the definition of complex or imprecise concepts', *International Journal of Man-Machine Studies* (8), 249–291.

Zadeh, L.A. (1979). 'Fuzzy sets and information granularity', in M.N. Gupta, R.K. Ragade & R.R. Yager (eds.), *Advances in fuzzy set theory and applications,* North-Holland Publishing.

Zebrowski, E.Jr. (1983). 'College science textbook publication: A look at the sociological mechanism', *Science Education* (67), 443–453.

Zeitoun, H.H. (1984). 'Teaching scientific analogies: A proposed model', *Research in Science and Technology Education* (2), 107–125.

AUTHOR INDEX

Abrams, E., 25, 138, 185, 198
Aleixandre, M.P., 62, 69, 185
Al-Kunifed, A., 139, 185
Alldredge, S., 161, 191
Altman, L.K., 42, 185
Alvermann, D.E., 170, 191
Ambron, J., 10, 11, 21, 185
American Association for the Advancement of Science (AAAS), 9, 32, 35, 97, 101, 185
Amlund, J.T., 153, 185
Anderson, C., 68, 177, 196
Anderson, C.W., 56, 62, 185
Anderson, D.K., 79, 197
Anderson, J.R., 84, 154, 185
Andre, T., 68, 199
Aristotle, 154, 185
Arons, A.B., 67, 185
Atrans, S., 28, 185
Ault, C.R., 169, 185
Ausubel, D.P., 7, 16, 30, 56, 80, 86, 136, 144, 185
Avrams, R., 59, 193
Baars, B.J., 8, 48, 162, 185
Barinaga, M., 95, 185
Barowy, B., 68, 191
Barrow, L.H., 73, 160, 194
Barufaldi, J.P., 169, 197
Basili, P.A., 68, 185
Beckwith, R., 164, 193
Beissner, K., 8, 10, 70, 143, 191
Berra, T.M., 178, 185
Beyerbach, B., 16, 185
Biological Sciences Curriculum Study, 180, 185
Bishop, B.A., 62, 185
Blakeslee, S., 68, 95, 111, 186
Bourne, L.E.Jr., 83, 84, 186
Boyes-Braem, P., 151, 196
Brachman, R.J., 6, 143, 159, 186
Bradford, D., 161, 186
Branaghan, R.J., 14, 186
Brewer, W.F., 67, 73, 82, 186
Brigham, F.J., 162, 186
Brinkman, F., 68, 187
Bronowski, J., 29, 186
Broudy, E., 178, 186
Brown, R., 59, 84, 186
Bruer, J.T., 36, 186
Brumby, M., 59, 61, 186
Burk, J., 11, 12, 21, 192
Buzon, B., 7, 12, 13, 21, 144, 162, 186
Buzon, T., 7, 12, 13, 21, 144, 162, 186

Bybee, R.W., 173, 196
Cain, S.E., 10, 186
Carey, S., 59, 186
Carpenter, T.P., 89, 190
Champagne, A., 58, 59, 186
Chase, W.G., 152, 186
Chi, M.T.H., 84, 186
Chiappetta, E.L., 111, 187
Chinn, C.A., 67, 73, 82, 186
Chmielewski, T.L., 155, 186
Chomsky, C., 68, 186
Christianson, R.G., 73, 160, 186
Christy, O.B., 171, 186
Clark, W.R., 49, 52, 186
Clement, J., 58, 67, 90, 186, 187
Clements, D.H., 10, 187
Cliburn, J.W.Jr., 8, 187
Cobern, W., 27, 64, 68, 187
Cohen, D.K., 9, 187
Collette, A.T., 111, 187
Collins, A.M., 58, 84, 143, 187
Cooke, N.M., 155, 187
Cooper, L.A., 153, 197
Corno, L., 40, 197
Coulson, R.L., 79, 156, 197
Craig, R., 44, 45, 187
Cretzinger, J.I., 171, 172, 187
Cross, D.R., 159, 192
Crovello, T.J., 10, 187
Cummins, C.L., 64, 189
Cushing, S., 106, 107, 187
Dagher, Z.R., 68, 187
Dalton, B., 68, 187
Dansereau, D.F., 155, 159, 186, 191, 192
Davis, R., 145, 156, 187
Dawkins, R., 102, 187
De Jong, O., 68, 187
de la Rocha, O., 84, 192
Demastes, S.S., 62, 64, 65, 187, 189
Dennett, D.C., 43, 46, 187
DiGisi, L.L., 170, 187
Dodson, B., 32, 46, 47, 190
Dominowski, R.L., 83, 84, 186
Donaldson, M., 10, 187
Doster, E.C., 61, 191
Driver, R., 10, 187
Dubay, J., 56, 185
Dubin, R., 80, 188
Duffield, J., 145, 190
Duit, R., 58, 195
Dunlap, D.W., 105, 145, 188, 190
Edmondson, K.M., 170, 188

201

Elliott, D.L., 168, 188
Ericsson, K.A., 59, 188
Evans, J.M., 10, 186
Everson, H.T., 156, 198
Eylon, B.-S., 176, 177, 188
Fairchild, 164, 188
Faletti, J., 7, 17, 22, 30, 154, 156, 188
Fegahli, A., 159, 188
Fellbaum, C., 164, 193
Feltovich, P.J., 79, 84, 156, 186, 197
Fennema, E., 89, 190
Fensham, P.J., 10, 70, 188, 199
Fisher, K.M., 7, 10, 17, 22, 30, 58, 59, 70, 73,
    79, 82, 96, 98, 99, 124, 144, 154–156, 159–
    161, 163, 180, 181, 186, 188–190, 193
Flavell, J.H., 91, 160, 189
Flick, L.B., 9, 189
Fowler, R.H., 164, 189
Freeman, K., 132, 189
Freiberger, P., 151, 193
Fremerman, S., 42, 189
French, R.M., 100, 189
Friedlander, B.P., 110, 189
Fuller, R.G., 67, 189
Furnas, G.W., 164, 188, 189
Fuson, K.C., 89, 190
Gaffney, J., 153, 185
Gagne, R.M., 78, 189
Gal-Choppin, K., 59, 197
Gallo, M., 138, 191
Gamas, W.S., 68, 190
Gardner, E.J., 39, 115, 189
Garrard, J.E., 10, 70, 188, 199
Garvie, L.A.J., 150, 189
Gentner, D., 58, 99, 145, 187, 189
Gerzog, W.A., 67, 195
Gilbert, J., 16, 59, 194, 195
Gill, R.T., 46, 79, 145, 189
Gittings, E.C., 161, 186
Glaser, R., 84, 186
Glass, G.V., 68, 190
Globerson, T., 10, 196
Glynn, S.M., 104, 189
Gomes, L.M., 10, 79, 156, 188, 195
Good, R., 62, 64, 65, 170, 187, 189, 197
Gordon, S.E., 19, 22, 46, 70, 79, 145, 189
Gordon, W.T., 100, 190
Gottfried, S.S., 168, 177, 190
Gould, S.J., 64, 102, 190
Gowin, D.B., 10, 15, 22, 111, 123, 144, 194
Grabiner, J.V., 172, 173, 190
Grabinger, R.S., 144, 145, 190, 193
Gravett, S.J., 56, 70, 190
Gray, W., 151, 196
Greenes, R.A., 164, 191
Griffard, P.B., 128, 190

Gross, D., 164, 193
Gross, P.R., 34, 190
Gullo, D.F., 10, 187
Gunstone, R.F., 58, 59, 74, 123, 186, 190, 199
Guzzetti, B.J., 68, 190
Hackney, M.W., 101, 190
Hafner, M.S., 64, 189
Halloun, I., 73, 190
Hamelin, D., 17, 138, 191
Hammond, N., 160, 195
Hanesian, H., 16, 30, 136, 185
Haslam, F., 56, 73, 190
Hegarty, M., 153, 191
Heller, J.I., 84, 148, 195
Hellmann, R.A., 171, 172, 190
Helms, H., 58, 190
Hendricks, P.L., 162, 186
Hestenes, D., 73, 190
Hettich, P.I., 111, 190
Hewson, M.G., 67, 190
Hewson, P.W., 67, 190, 195
Hiebert, J., 89, 190
Hoagland, M., 32, 46, 47, 190
Hoffman, R.P., 154, 190
Hofmann, J., 159, 191
Holden, C., 127, 128, 191
Holley, C.D., 155, 191
Holliday, W.G., 170, 191
Horn, R.E., 7, 146, 191
Horton, P., 17, 138, 139, 191
Horwitz, P., 68, 191
Howard Hughes Medical Institute, 30, 191
Hoz, R., 16, 191
Hyerle, D., 111, 191
Jackson, D.F., 40, 61, 191, 197
Jacobson, M.J., 156, 191, 197
Jay, M., 161, 191
Jeannerod, M., 49, 191
Jegede, O., 128, 194
Johnson, D.W., 151, 163, 191, 196
Johnson, M., 90, 192
Johnson, R., 163, 191
Jonassen, D.H., 8, 10, 70, 143, 164, 191
Karplus, R., 67, 189
Kelly, G.A., 16, 43, 87, 191
Kibby, M.R., 10, 17, 144, 188
Kliebard, H.M., 191
Klopfer, L., 58, 59, 186
Knol, W.E., 68, 198
Komorowski, H.J., 164, 191
Kourik, R., 29, 191
Kozhenikov, M., 153, 191
Krajcik, J., 16, 170, 197
Kuhn, D., 160, 191
Kuhn, T.S., 81, 165, 192
Kulhavy, R.W., 153, 185

Kutcka, S.M., 162, 186
Kyle, W.C.Jr., 168, 177, 190
Lakoff, G., 90, 100, 151, 192
Lambiotte, J.G., 159, 192
Langer, E.J., 2, 10, 78, 91, 123, 147, 156, 192
Langer, S., 156, 192
Lanzing, J.W.A., 135, 192
Larkin, J., 123, 148, 192, 195
Lave, J., 84, 192, 196
Lavoie, D.R., 137, 192
Lawson, A., 67, 189
Lehman, F., 143, 157, 192
Levesque, H.J., 6, 143, 159, 186
Levy, G.B., 121, 192
Lieberman, P., 95, 192
Linder, C.J., 68, 192
Linn, M.C., 176, 177, 188
Lipson, J.I., 7, 17, 22, 27, 59, 188
Liu, M., 153, 159, 192
Lloyd, B.B., 151, 196
Lloyd, C.V., 168, 170, 192
LNR Research Group, 10, 84, 194
Locke, J., 55, 192
Loftus, E.F., 83, 84, 143, 186, 187
Logan, J., 162, 192
Longo, P.J., 19, 20, 22, 192
Lovitt, Z., 11, 12, 21, 192
Luoma-Overstreet, K., 154, 192
MacHarg, R.A., 169, 199
Macnamara, J., 159, 192
Magner, L.N., 39, 192
Markman, E.M., 111, 192
Marra, R.M., 156, 192
Mason, L., 68, 192
Mastrilli, T.M., 104, 193
Mayer, R.E., 152, 153, 191, 193
Mayr, E., 28, 29, 60, 193
Mazlish, B., 29, 186
McAleese, R., 70, 144, 145, 193
McClelland, J.L., 48, 196
McComas, W.F., 57, 193
McConney, A., 17, 138, 191
McDonald, J.E., 155, 187
McKnight, C.C., 88, 196
McNeill, D., 84, 151, 165, 186, 193
Mead, P.L.R., 68, 187
Meadows, L., 61, 191
Meighan, M., 98, 193
Merkin, A., 161, 186
Mervis, C., 151, 196
Miles, M.B., 143, 199
Miller, G.A., 164, 193
Miller, J.G., 2, 29, 47–52, 193
Miller, K., 164, 193
Miller, P.D., 172, 173, 190

Mintzes, J.J., 56, 58, 68, 83, 91, 123, 127, 136, 137, 139, 193, 197–199
Moody, D.E., 64, 65, 173–175, 193
Moore, J.A., 29, 193
Morgan, R.L., 161, 186
Morine-Dershimer, G., 16, 68, 193
Morocco, C.C., 68, 187
Murray, B., 78, 193
Murray, H., 89, 190
Murtaugh, M., 84, 192
Muschamp, H.T., 32, 193
Musonda, D., 16, 194
Nachmias, R., 59, 193
National Research Council, 9, 89, 177, 178, 193
National Science Board Commission on Precollege Education in Mathematics, Science and Technology, 89, 193
National Science Board Task Committee, 89, 194
National Science Foundation, 89, 194
Nelkin, D., 173, 178, 194
Nichols, M.S., 123, 194
Nix, D., 156, 197
Nobles, C.S., 123, 194
Norman, D.A., 10, 59, 80, 84, 91, 194, 196
Novak, J.D., 10, 15, 16, 22, 30, 56, 58, 59, 83, 91, 114, 123, 127–129, 136, 139, 144, 185, 190, 193–195, 199
Nussinovitz, R., 59, 197
Obourn, E.S., 10, 199
Odom, A.L., 73, 160, 194
Okebukola, P., 17, 128, 169, 194
Olivier, A., 89, 190
Ortony, A., 23, 100, 194, 196
Osborne, J., 59, 194
Otto, J.H., 174, 194
Pagels, H.R., 75, 156, 194
Paivio, A., 120, 194
Papert, S., 10, 194
Pask, G., 6, 10, 194, 195
Patterson, H.A., 7, 17, 22, 188
PDP Research Group, 48, 196
Pea, R.D., 10, 195
Peebles, P., 65, 187
Peirce, C.S., 152, 195
Perkins, D.N., 10, 144, 161, 162, 195
Peters, F., 161, 191
Pfundt, H., 58, 195
Pines, A., 58, 59, 70, 145, 164, 195, 199
Pinker, S., 68, 157, 162, 195
Poltrock, 164, 188
Pope, M., 16, 58, 195
Posner, G.J., 59, 61, 67, 73, 146, 195, 197
Quillian, M.R., 7, 143, 195
Raizen, S.A., 88, 196

Reader, W., 160, 195
Reif, F., 84, 148, 195
Reiger, C., 148, 195
Reisbeck, C.K., 148, 195
Resnick, L.B., 10, 92, 195
Reynolds, L., 120, 195
Reynolds, S.B., 159, 192
Richens, R.H., 157, 196
Rico, G.L., 10, 196
Robinson, A., 136, 196
Rodin, E.Y., 143, 192
Rogoff, B., 84, 196
Rosch, E., 151, 196
Rosenberg, A., 29, 196
Rosenthal, D.B., 173, 196
Rosenthal, J.W., 99, 145, 196
Roth, K., 177, 196
Rowell, R., 7, 128, 196
Ruiz-Primo, M.A., 139, 196
Rumelhart, D.E., 10, 23, 48, 80, 84, 91, 194, 196
Rushworth, P., 10, 187
Sagan, C., 39, 196
Sakurai, J., 39, 196
Saljo, R., 68, 198
Salomon, G., 10, 195, 196
Sanford, J.P., 68, 185
Schmidt, W.H., 88, 196
Schnepps, M., 1, 55, 71, 72, 82, 196
Schuette, E.E., 162, 186
Schvaneveldt, R.W., 14, 21, 145, 155, 196
Senn, G., 17, 138, 191
Settlage, J.Jr., 62, 187
Shavelson, R.J., 139, 196
Shea, C., 105, 196
Sheldon, T.H., 56, 185
Shepard, R.N., 153, 197
Shrobe, H., 145, 187
Shymansky, J.A., 170, 197
Simmonds, D., 120, 195
Simon, H.A., 51, 59, 123, 152, 186, 188, 192, 197
Sims, V.K., 152, 193
Skoog, G., 173, 197
Slavin, R., 163, 197
Smith, J., 16, 185
Smith, K., 163, 191
Snow, R.E., 40, 197
Snyder, T.E., 68, 190
Songer, C.J., 68, 197
Sowa, J., 10, 143, 197
Spence, M., 104, 105, 197
Spiegel, G.F., 169, 197
Spiro, R.J., 79, 156, 191, 197
Spring, C., 7, 17, 22, 188
Squires, A., 10, 187

Starr, M., 16, 170, 197
Starratt, S., 98, 193
Stavy, R., 59, 193
Stepans, J., 59, 197
Stevens, S.S., 113, 197
Stevenson, H.W., 85, 86, 197
Stewart, J., 7, 128, 197
Stigler, J., 85, 86, 197
Stoddart, T., 68, 197
Stofflett, R.T., 68, 197
Strike, K.A., 61, 67, 146, 195, 197
Swackhamer, G., 73, 190
Swart, E., 56, 70, 190
Szolovits, P., 145, 187
Tamir, P., 16, 59, 191, 197
Taveggia, T.C., 80, 188
Thomson, E.A., 139, 140, 197
Thorley, N.R., 68, 197
Thornton, R., 7, 17, 22, 188
Thro, M.P., 10, 145, 198
Tillema, H.H., 68, 198
TIMSS, 88, 198
Tivnan, T., 68, 187
Tobias, S., 156, 198
Tomer, Y., 16, 191
Towle, A., 174, 175, 194, 198
Treagust, D.F., 56, 73, 190
Triangle Coalition for Science and Technology Education, 35, 198
Troost, C.J., 172, 198
Trowbridge, J.E., 64, 127, 137–140, 189, 198
Tufte, E.R., 113, 198
Tyson-Bernstein, H., 178, 198
Van Kirk, J., 7, 128, 197
VanKirk, J., 59, 195
von Glasersfeld, E., 58, 87, 198
Vosniadou, S., 68, 198
Vygotsky, L.S., 144, 198
Wade, R.C., 68, 198
Waldholz, M., 77, 198
Walker, D.F., 168, 198
Wallace, J., 123, 137, 139, 198
Wan, E., 98, 193
Wandersee, J.H., 7, 15, 16, 21, 25, 26, 28, 30, 35, 56, 58, 64, 74, 83, 91, 101, 111, 113, 114, 115, 120, 121, 127, 128, 130, 134, 136–140, 169, 185, 189, 190, 193, 194, 198, 199
Wang, S., 164, 191
Watson, J., 82, 199
Wearne, D., 89, 190
Weinberg, S.L., 178, 199
Weiner, J., 81, 199
Weiss, I.R., 168, 177, 199
Weitzman, E.A., 143, 199
Wells, M., 73, 190
Welschselgartner, B., 159, 191

West, L.H.T., 10, 58, 70, 145, 164, 188, 199
White, R., 74, 123, 199
Wilford, J.N., 130, 131, 199
Willerman, M., 169, 199
Willett, J.B., 170, 187
Williams, J.L., 164, 189
Wilson, E.O., 68, 101, 171, 185, 199
Windschitl, M., 68, 199
Winner, E., 102, 103, 199
Wittrock, M.C., 10, 16, 199
Wood, T., 61, 191
Woodburn, J.H., 10, 199
Wood-Robinson, V., 10, 187
Woods, A., 17, 138, 191
Woodward, A., 168, 188
Yacci, M., 8, 10, 70, 143, 191
Yore, L.D., 168, 170, 177, 191, 197, 199
Zadeh, L.A., 151, 199, 200
Zebrowski, E.Jr., 178, 200
Zeitoun, H.H., 104, 200

SUBJECT INDEX

Ability, acquired versus innate, 85, 86
Academic rigor, 43
Accommodation, 91
Accretion, 32, 91
Affordances, 124
Air-ground communications, 106, 107
Algorithms, 23
American Association for the Advancement of Science (AAAS), 9, 32, 35, 97, 101, 127
Analog, 30, 100, 102
Analogies, 31, 68, 90, 100–104, 108, 137
Analogy teaching models, 104
Analysis
    Cost-benefit, 96
    Curriculum, 16, 171, 172
    "Paralysis of analysis", 43, 140
    Plane of, 52
    Process, 99
    Supporting student, 80, 87, 144, 156, 161
Anatomy, 18, 169, 171
Anomalous data, responses to, 67
Aptitude, 85
Artificial intelligence (AI), 6, 7, 22, 143, 164
Assessment, 18, 74, 138, 160, 163,
Assimilation, 68, 91, 96, 98, 136
Associations, 5, 8, 10, 21, 22, 59, 157
Ausubelian learning theory, 7, 16, 30, 56, 80, 86, 113, 136, 144
Automaticity, 78, 84, 91
Basic Biology Corpus, 98, 99
"Believing is seeing", 82
Benchmarks, 32, 54, 73, 137
Bidirectional links, 5, 18, 19, 22, 124, 150, 162
"Big picture", 45, 46, 54, 79
Biological denial, 33
Biological literacy, see *Literacy, scientific*
Biology, see also other categories in index
    History of, 28–30, 34, 39, 57, 82, 101, 171, 176, 183
    Jargon, 96
    Knowledge, 25–33, 44–47, 51–53, 70, 130, 145, 154, 163, 183, see also *Mapping biology knowledge*
    Maps, see *Mapping biology knowledge*
    Structural map of, 174, 175
Bottom-up instruction, 46
Breadth (coverage), 30, 35, 88, 173
Bridging experiences, 90, see also *Metaphors* and *Analogies*
Broadcasting (to all the modules in the brain), 8, 48, 162
BSCS series, 173
Buffer inventories, 49
Carbon dioxide, 55, 56, 58, 74, 119
Cardiff Giant hoax, 32
Cartography of cognition, 30, 130
Categories (idea organization), 7, 15, 51, 91, 111, 129, 150, 151, 163
Channel (as part of living systems), 48, 51
Chromosomes, 49, 82, 83, 150, 152

Chronology, 53, 99, 105, 114, 117
Chunks (of information), 151, 152
Circle diagram, see *Concept circle diagrams*
Circle template, 113
Cluster maps, 10–14, 21
C-Map (software), 135
Codified knowledge, 90
Cognitive, 180, 184, see also *Metacognition*
    Art, 113
    Conflict, 56–61, 68, 73–75, 81
    Distance, 61, 65, 66
    Errors, 59
    Flexibility theory, 79, 156, see also *Flexibility*
    Framework, 53, 82, 91, 103
    Infrastructure, 46, 47
    Research, 10, see also *Situated cognition*
    Science, 7, 36, 103, 143, 164
    Skills, 8, 10, 84–87, 155, 159, 163, 165
    Snapshot, see *Snapshot, conceptual*
    Structures, see *Knowledge structures*
    Variables, 80
Cognitively guided instruction (CGI), 89
Concept
    Alternative, see *Conceptions, alternative*
    Circle diagram, see *Concept circle diagrams*
    Defined, 110, 129, 131, 141
    Embedded, see *Embedded concepts*
    Force Inventory Test, 73
    Introduction of, 96, 110
    Knowledge-map element, 5, 21, 22, 69, 116, 123, 128, 131–134, 137, 141, 151–153
    List, 99, 102
    Map, see *Concept maps*
    Protoconcepts, 74
    Seed concepts, 137
Concept circle diagrams (CCDs), 7, 15, 16, 110–126, 179
    Graphic complexity of, 123
Conceptions, see also *Misconceptions* and *Preconceptions*
    Alternative, 36, 56, 57, 61–64, 69, 73, 74, 83,
    Naive, 56–59, 73, 77, 82, 83
Concept maps, see also *Mapping*
    Cartography and, 130, 131
    Coconstructed maps, 139
    Conventions in, 133–137, 141
    Description of, 128, 141
    Evaluating, 137, 141, 142
    Graphic effectiveness of, 134
    Iconic, 139
    Invention and theoretical bases of, 7, 15, 16, 127, 128, 135, 136, 141
    Novakian, 7, 131, 135, 140, 141
    Research on, 17, 137–140, 142, 164, 169
    Software, 7, 15, 17, 135, 155
    Textbooks and, 169, 170, 178, 179
    Types of and related techniques, 18, 56, 69, 123, 124, 129, 139
    Uses of, 16, 17, 127, 135, 141, 155, 164, 169, 183
Concept nodes, 18, 143
Conceptual change, 16, 57, 61, 62, 65–75, 124, 138, 164

Conceptual conflict, see *Cognitive conflict*
Conceptual ecology, 59, 75
Conceptual graph, 18, 22, 79
Conceptual histories, 130
Cone of specificity, 99
Conflicting beliefs, see *Cognitive conflict*
Conflicting forces (in textbook industry), 178
Connectedness, 146, 181
Connections
    Among ideas, 6, 7, 16, 53, 54, 84, 130, 141, 147, see also *Relations*
    To prior knowledge, 85, 87, 93
Constructivism, 16, 57, 58, 67, 80, 87, 90, 100, 135, see also *Human constructivism*
Constructs, 23, 27, 116, 131, 134, 141
Content drives relations, 154
Content knowledge, 46, 58, 79, 145
Continuum, strength-of-relationship, 100, 129
Conventional teaching strategies, 57, 79, 169, 170
Conventions, see under *Concept maps*
Conversation, 6, 7, 85, 87, 103, 134
Cost-benefit analysis, 96
Coverage (of content), 74, 88, 172, 173, 177
Creator's design, 60
Criterion-referenced checklist, 137
Cross-language learning, 159
Cross-links, 131, 134, 135, 137, 141, 142
Curriculum, 16, 33–35, 67, 68, 74, 88, 89, 92, 111, 138, 167, 168, 170–173, 177–179, 183
Curriculum analysis, 16
Cybernetic knowledge mapping, 6
Cybernetic theory, 6
Detail recall, 11, 25, 43–46, 53, 54, 79, 83, 112, see also *Memorization*
Diagnostic feedback, 145
Diagram conventions, 116, see also *Concept circle diagrams*
Dialogue, absence of, 86
Differentiation, see *Discrimination*
Discrimination (among ideas), 7–9, 82, 83, 145, 146, 150, 151, 155, see also *Distinctions*
Disparate viewpoints, 81
Display labels, 97
Distinctions, 45, 79, 87, 95, 151, 170
Ditto sheets, 86
Double scanning, 121
Dry log, 55
Dual-coding theory, 120, 121, 152, 153
Dynamic process, 47, 53, 149
Eco-cultural niche, 105
Economy-of-information rule, 43
Edge detection, 120
Educational reform, see *Reform movement*
Effects of/effects with information processing technologies, 161, 162, 173
Electron transport chain, 84
Electronic technologies, 10
Embedded concepts, 59, 74, 147, 148, 155, see also *Hierarchies*
Emergent properties, 29
Encoding, 49, 51, 81, 83–85, 93, 152
Encyclopedia of ideas, 164
Epistemological meta-knowing, 160
Errors, 59, 65–67, 83–85, 106, 108

Essentialism, 60
Euler's circles, 114, 115
Evidence, 40, 57, 73, 75, 77, 81
Evolution, 3, 29, 33, 60–65, 70, 101, 131, 163, 170–176, 180–182
Examples, 31, 46, 61, 68, 78, 104, 129–131, 134–137, 141, 142
Expectation-generator, 2, 43
Experts, see under *Knowledge*
Eye-track pattern, 120
Factoids, 44, 53
Facts, 17, 40, 44–46, 53, 75, 79–81, 160, see also *Evidence*
Fixedness or rigidity of thinking, 80
Flexibility (in thinking), 31, 78–80, 90, 93, 141, 156, 162, 177, 181
Flows, temporal, 147–149, 163
Flynn effect, 105
Four Cs, 163
Frames (in semantic newtorks), 17, 18, 151–153, 157, 159, 162–164
Future-focused, 43
Fuzzy boundaries, 117
Fuzzy ideas, 8, 29, 31, 83, 90, 151
Fuzzy set theory, 151
General principles (of biology), 32, 46, 47, 52, 53
Generalizability of explanations, 29
Generalizations, cross-level, 51, 52
Genetics, 17, 25, 29, 169
Geographic mapping, 23
Group work, 10, 74, 79–81, 83, 86–90, 144, 162–164
Guided discovery, 90, 159
Hierarchies
    Conceptual, 43
    Construction of, 15, 53, 84, 131–134, 141, 163
    Embedded, 147
    Levels of, 52, 114, 124, 131, 148–150
    Limited or inappropriate propositional, 56
    Linkable, 133
    Structure, 128, 137
Higher order thinking, 91–94, 116, 131
Historical baggage, 82
Holistic images, 111
Human constructivist learning theory, 114
Imposing meaning, 77, 92
Inclination to question, 93, see also *Openness* and *Flexibility*
Inclusive-exclusive relationships, 110, 111, 125
Inert, see under *Knowledge*
Informavores, 43
Innate pleasure of learning, 79
Inquiry learning, 9, 10, 36, 57, 61, 67, 73, 75, 89, 183
Insight, 64, 65, 67, 74, 75, 82
Inspiration (software), 7, 8, 135
Intellectual endeavor, 61
Intellectual prostheses, 100
Intensities of information inputs, 51
Intention and belief, 102, 103, see also *Misconceptions*
Interconnectivity, 9, 146, 147, 164
Interpretations, 68, 82, 83, 85, 92, 93, 122, 123
Knowing Biology, 25, 31, 39
Knowledge

Applying, 84
Ease and fluidity of, 91
Capture, 127
Cognitive, 87
Consolidation, 90
Construction, 9, 16, 58, 70, 87, 94, 144, 163
Core, 147
Effortful, 87, 90, 92
Expert, 8, 9, 19, 22, 25, 30, 31, 84, 85, 91, 155
Embedded within context, 84
Experiential, 90, 94
Explicitly organized, 90
Inert, 86, 130, 156
Mapping, 5–10, 18, 20–23, 43, 47, 70, 74, 144, 159, 160, 165
Representation, 5, 6, 15, 143, 146, 160, 165, 179
Semantic, 94
Structural, 8, 23, 70, 164
Structures, 8, 17, 23, 43–47, 52, 53, 57, 65, 69, 91, 110, 125, 133, 145–151, 156
Learning for understanding, 10, 85
Learning without thinking, 78
Lecture, 73, 83, 86, 89, 98, 138, 160, 161
Legitimate scientific question, 42, see also *Evidence*
Less is more, 88, 89
Linguistic consistency, 99
Literacy, scientific, 32, 96, 111, 165
Living things, 14, 28, 29, 60, 64
Macromap, 134
Main ideas, 111, 147, 148, 155
Mapping, see also *Concept maps*
Benefits of, 8–10, 32, 37, 56, 68–70, 85, 144, 159
Biology knowledge, 30, 31, 36, 37, 47, 52, 105, 106
Geographic, 23
Knowledge, see *Knowledge mapping*
Software, 8
Strategies, 7, 10, 23, 94, 129, 165
Types introduced, 10–22
Matrix of meaning, 75, 157, 159
Meaningful learning
Described, 80–85, 96, 170
Higher order thinking and, 91–94
Mapping as a tool for, 7, 9, 17, 36, 116, 135, 136, 144, 160, 164
Mindful learning and, 53, 68, 77, 78, 93
Personally, 79
Reform movement and, 9
Rote learning versus, 30, 53, 74, 75, 86, 87 (chart), 144
Meiosis, 19, 83, 141
Memory
Long-term, 7, 17, 23, 44, 49, 70, 85, 93, 100
Short-term or working, 8, 23, 48, 70, 117, 134, 144, 162
Memorization, 43–45, 79, 86, 94, 144, 177, see also *Rote learning*
Memory extender, 70
Mental model, 67, 70, 83, 88, 90, 145
Mental simulation, 90
Meta-analysis, 17, 68, 138, 142
Metacognition, 70, 74, 113, 156, 160
Metacognitive skills, 8, 10, 85, 87, 128, 141, 161, 165, 170, see also *Cognitive skills*

212

Meta-learning, see *Metacognition*
Metaphor of death, 98
Metaphors, 10, 23, 28, 90, 100–104, 108
Metastrategic skill, 160
Methodology, instructional, 92
Micromap, 134
Miller Analogies Test (MAT), 104, 105
Mind maps, 6, 7, 12, 14, 19, 21, 162
Mindful Learning, see under *Meaningful learning*
Mindless learning, see *Memorization*
Minds-on learning, 170
Misconceptions, 55–59, 64–68, 70, 73–75, 83, 91, see also *Conceptions*
Misinterpretation, see *Misconceptions*
Motivation, 60, 80, 93, 184
Multiple-choice tests, 59, 73, 75, 86, 104, 163
Multiple criteria, 92
Multiple layers of information, 23
Multiple perspectives, 78, 79, 93, 123, 141, 147, 168
Multiple solutions, 91
Museum, 97
Natural selection, 29, 60, 61, 64, 69
"Need" questions, 64
Neo-Darwinian evolution, 60
Net
    Building, 160–163, 181
    (Semantic network), 124, 146–150
    Shown in figures, 147, 149, 150, 152, 153, 158
    Subsystem in living things, 48
Networks of ideas, 17, 31
Neuroscience, 36
Nonliteral language, 102
Noun concepts, 129, 130
Novakian concept maps, see under *Concept maps*
Novelty (importance for learning), 78
Nuanced judgment, 92
Off-loading, 43
Openness, see *Flexibility*
Overlearning, 78
Pack rat, 45
Parade of diagrams, 118
Parentese, 95
Parsimony, 146, 150
Pathfinder software, 14, 155
Pattern recognition, 86, 105, 129, 152
Pedagogical content knowledge, 58, see also *Knowledge*
Perception, 57, 75, 81, 83, 85, 93, 100, 114, 120, 121
Perception-based rules, 118
Persistence
    In learning efforts, 94
    Of ideas, 56, 62, 70, see also *Misconceptions*
Personality determination, 39–42
Photosynthesis, 55, 58, 66, 70–74, 99, 131, 163
Planning, 16, 138, 160, 177, 179
Popular press publications, 101
Practice, importance of, 8, 18, 84, 86, 87, 91, 94, 97, 98, 129, 138, 156
Preconceptions, 56, 58, 62, 64, 67, see also *Misconceptions*

Predator, 83
Prediction, 90
Prescientific ideas, 56, see also *Misconceptions*
Priming, 143
Prior knowledge, 11, 57, 58, 78, 85, 87, 91, 93, 100, 129, 164, 170, 184
Probing questions, 90
Problem solving, 8, 9, 79, 86, 89, 145, 148
Process words, 98
Processing capacity, 117
Project 2061, 9, 35
Propositional summary, 117, 120
Protoconcepts, see under *Concepts*
Psychologically sized, 113
Public controversies, 171
Public understanding of science, 131
Quantifying meaning, 51
Reality, approximation of, 87
Reduction and holism, 101
Reflection, 9, 30, 37, 70, 87, 110, 125, 128, 144, 148, 156, 160, see also *Metacognition*
Reform movement, 9, 10, 35, 78
Relations, 5, 8, 15, 18–23, 30, 43, 45, 51, 60, 111, 130, 145–156, 161
    Ubiquitous, 155
Relationship, 5, 36, 69, 100, 104, 114, 116, 117, 123, 168, 176
Reserve final judgment, 42
Resistance to change, 59, 65, see also *Misconceptions*
Restructuring, 65, 91
Rethinking ideas, see *Revision*
Retrieval, 84, see also *Memory*
Revision, 37, 70, 79, 87, 156, 160
Rigidity, 80, see also *Resistance to change*
Rote learning, 7, 43, 79, 86, 94, 110, 123, 164, see also *Memorization*
Rule of thumb, 137, 141
Santa Claus, 61
Schema, 23, 152
Science, borderline or fringe areas of, 42
Scopes trial, 172, 173
Second-language learners, 145
Self-regulation, 92
Semantic center, 99
Semantic networks
    Described in tables, 22, 123, 124
    Examples in figures, 18, 146, 153
    Hierarchies in, 150
    Research on, 143
    Semantic network theory, 7, 143, 144, 164
    Semantic networking, 18, 85, 107, 144, 160, 161
    SemNet® as tool for, 17, 18, 69, 107, 143–146, 151, 153–156, 159–165
Semantic structure, 8, 143
SemNet® (software), 7, 17, 18, 69, 107, 123, 124, 143–165, 180–183
Sense-making, 57, 85
Situated cognition, 84
Skeletal ideas, 156, 157
Skill development, 88, 91
Snapshot, conceptual, 23, 70, 122–126
Social interaction, 87, see also *Group work*

Space-age fundamentalism, 173
Spatial navigators, 53
Spreading activation, 84, 143, 164
Structure of the subject matter, 173, 176
Subconscious, 2, 8, 60, 108, 162
Superordinate concept, 131, 134, 141
Symbolic descriptions, 90
Synergistic, 87, 169
System of systems, 47
Target, 78, 100, 102
Task-specific organizers, 111
Teachers, 9, 11, 16, 30, 34–36, 43–48, 53–61, 68, 73–75, 79–81, ...
Teaching science for understanding, see *Inquiry learning*
Telescoping, 16, 116–118, 121, 124
Template paper (for concept maps), 139, 140
Terminology, 96, 98
Terms
    Choosing for biology, 96–99
    Multiple meanings for, 82, 83
    Relations between, 14, 174, 179, 180, 183
    Secondary layer of specialized terms, 97
    Textbooks and, 45, 174–176, 179
    Too precise, 106, 107
Textbooks
    Adoption committees, 178, 179, 183
    History of, 171–176
    Inappropriate emphases, 53, 83, 168, 170–176, 183
    Influence of, 167, 168, 177
    Outdated, 36
    Publishers, 178, 179, 183
    Research on use, 168–176, 183
    Term-laden, 45, 183
Time-on-task, 122
Tools
    Cognitive, 179
    Graphic representation, 168, 184
    Mapping, 69, 70, 75
    Metacognitive, 110, 113, 116, 123–125, 128, 141, 183
    Stand-alone evaluation, 123
Top-down instruction, 46
Transmission instruction, see *Lecture*
Tuning, 91
Two cultures (scientists and science educators), 9, 33–36
Uncertainty, 92
Understanding
    Benchmarks and, 54
    Biological, 30–34, 43–45, 52, 57, 73–75, 78, 81–83, 88, 105, 122–125, 130, 183
    Conceptual, 16, 27, 67, 68, 96, 100, 112, 121
    Meaningful, 7, 78–81, 86, 89, 93, 94, 110,129, 144, 164, see also *Meaningful learning*
    Misconceptions and, 57–61, 75
    Model of learning and, 28
    Novice to expert, 30
    Teaching for, 9, 10, 85, 93, see also *Inquiry learning*
    Textbooks and, 168–171
    Wisdom as, 57, 141
Unidirectional links, 5, 18, 19, 22

Venn diagrams, 114
Verbal horizons, 102
Visual dissonance, 114
Visual distillation effect, 112
Visual Thinking Network, 19, 20
Webs, 10–14, 21, 64, 129
Willingness to question, 80, see also *Flexibility*
Wisdom, see under *Understanding*
Worldview, 27, 28, 65, 101

Science & Technology Education Library

Series editor: Ken Tobin, *University of Pennsylvania, Philadelphia, USA*

Publications

1. W.-M. Roth: *Authentic School Science.* Knowing and Learning in Open-Inquiry Science Laboratories. 1995 ISBN 0-7923-3088-9; Pb: 0-7923-3307-1
2. L.H. Parker, L.J. Rennie and B.J. Fraser (eds.): *Gender, Science and Mathematics.* Shortening the Shadow. 1996 ISBN 0-7923-3535-X; Pb: 0-7923-3582-1
3. W.-M. Roth: *Designing Communities.* 1997
    ISBN 0-7923-4703-X; Pb: 0-7923-4704-8
4. W.W. Cobern (ed.): *Socio-Cultural Perspectives on Science Education.* An International Dialogue. 1998 ISBN 0-7923-4987-3; Pb: 0-7923-4988-1
5. W.F. McComas (ed.): *The Nature of Science in Science Education.* Rationales and Strategies. 1998 ISBN 0-7923-5080-4
6. J. Gess-Newsome and N.C. Lederman (eds.): *Examining Pedagogical Content Knowledge.* The Construct and its Implications for Science Education. 1999
    ISBN 0-7923-5903-8
7. J. Wallace and W. Louden: *Teacher's Learning.* Stories of Science Education. 2000
    ISBN 0-7923-6259-4; Pb: 0-7923-6260-8
8. D. Shorrocks-Taylor and E.W. Jenkins (eds.): *Learning from Others.* International Comparisons in Education. 2000 ISBN 0-7923-6343-4
9. W.W. Cobern: *Everyday Thoughts about Nature.* A Worldview Investigation of Important Concepts Students Use to Make Sense of Nature with Specific Attention to Science. 2000 ISBN 0-7923-6344-2; Pb: 0-7923-6345-0
10. S.K. Abell (ed.): *Science Teacher Education.* An International Perspective. 2000
    ISBN 0-7923-6455-4
11. K.M. Fisher, J.H. Wandersee and D.E. Moody: *Mapping Biology Knowledge.* 2000
    ISBN 0-7923-6575-5

KLUWER ACADEMIC PUBLISHERS – DORDRECHT / BOSTON / LONDON

Printed in the United States
1211100001B/26

9 781402 002731